Living Through the End of Nature

Living Through the End of Nature

The Future of American Environmentalism

Paul Wapner

The MIT Press
Cambridge, Massachusetts
London, England

For information about special quantity discounts, please send email to <special_sales@mitpress.mit.edu>.

This book was set in Sabon by the MIT Press.
Printed on recycled paper and bound in the United States of America.

Library of Congress Cataloging-in-Publication Data

Wapner, Paul Kevin.
Living through the end of nature : the future of American environmentalism / Paul Wapner.
    p.   cm.
Includes bibliographical references and index.
ISBN 978-0-262-01415-1 (hardcover : alk. paper)
1. Environmentalism—United States.   2. Conservation of natural resources—United States.   I. Title.
GE197.W37 2010
333.720973—dc22

                                                        2009036083

10   9   8   7   6   5   4   3   2

To Diane, Eliza, and Zeke

for everything

Contents

Acknowledgments

One of the main points of *Living Through the End of Nature* is that life is interdependent. The earth is a swirling mixture of plant, animal, and mineral with no entity able to exist without others. In writing this book, I have been fortunate to experience this in a concrete way. Not only has the sun shone down on me every day and the earth provided abundant sustenance; I have also lived and worked among concerned colleagues, inspiring students, dear friends, and a loving family. Acknowledging these connections is one of the deepest joys in completing the manuscript.

First, I would like to thank my colleagues in the School of International Service at American University. Dean Louis Goodman has long provided support and enthusiasm for my work, and through many conversations, helped clarify the arguments of the book. Simon Nicholson and Judy Shapiro each assisted me at critical junctures, and supplied sustained intellectual engagement. Leah Baker, Eve Bratman, Ritodhi Chakraborty, Benjamin Goldstein, Brendan Havenar-Daughton, Rongkun Liu, Antone Neugass, Marysia Szymkowiak, Rachna Toshniwal, Bonnie Washick, and Deidre Zoll offered valuable research assistance, for which I am particularly grateful. I am especially appreciative of my students. American University tends to attract people committed to making the world a better place. Whatever insights this book may have, many of them emerged through conversation and engagement with American University students.

Outside of American University, I have flourished in the company of scholars and others who have been caring about the earth in the most meaningful ways. Much of the book's argument crystallized while backpacking with Leslie Thiele and his many suggestions have improved the manuscript. Les has always provided intellectual companionship, wise counsel and lots of laughs. The book has also benefited from long country walks and conversations with Terry Post, conference discussions with Michael Maniates and Thomas Princen, and late night chats with Dirksen Bauman about the nature of nature. I presented aspects of the argument in seminars at Bowdoin College, University of Maryland, Rutgers University, the Lama Foundation, and the Center for Contemplative Mind in Society. I learned a lot from and am thankful for these opportunities. I am particularly grateful to Peter Dauvergne. Peter holds the record for reading the manuscript multiple times, encouraging me in moments of doubt, and generously offering incisive suggestions for improving the book's overall quality. Clay Morgan, at the MIT Press, read through the entire text with a keen editor's eye and expressed enthusiasm for the project in ways that have meant a lot to me.

From a different corner of my life, dear friends provided meaningful support as we lived together through the years in which this book was written. I would especially like to thank Dirksen Bauman, Joanna Bottaro, Sheila and Peter Blake, Kristin Dahl, Hilal Elver, Richard Falk, Sue Katz-Miller, Paul Miller, Robert Nelson, Mitchell Ratner, and Nicole Salimbene for the many conversations about nature as well as their friendship.

As always, my parents, Elinor and Morton, and siblings, Howard and Susan, have given me unwavering support and endless love. I thank them from the bottom of my heart.

This book is dedicated to my family. Eliza, Zeke, and Diane are the stars that shine daily for me and make life more than worth living. I suppose it is inappropriate to thank them for being who they are, but I sure appreciate them constituting my life.

Note on Photographs and Sources

Nicole Salimbene, a visual artist living in Takoma Park, Maryland, generously provided the photographs that appear before each chapter. Nicole works in a variety of media to explore themes of intimacy, political voice, and devotion. The photographs in the following pages are part of a study that examines the constructed quality of nature, and the place of memory and reverence in our engagement with the natural world. I am extremely grateful to Nicole for allowing her work to be shown in the context of the book and for inspiring me to think visually about the end of nature.

Zeke Wapner, my son, provided the cover photograph. The image captures his interpretation of what it means to live amidst profound change in both the human and natural worlds. I very much appreciate Zeke's keen photographic eye and his engagement with this book.

I have written the text in a way that I hope speaks to everyone. The challenges of living through the end of nature are not the sole domain of professional environmentalists or scholars of environmental studies, but demand engagement from all of us. With this in mind, I have tried to avoid academic jargon and to minimize the number of references. Citations appearing in the text refer only to essential works and those that most readers may be inclined to consult. The limited number of references should not hide the fact that I am indebted to countless scholars, journalists, and others for their insights into contemporary environmental affairs. This book is a product of many people—even if many of their names escape its pages.

Living Through the End of Nature

"Reweaving the Path," Lama, New Mexico

1

Introduction

If you stand right fronting and face to face to a fact, you will see the sun glimmer on both its surfaces, as if it were a cimeter, and feel its sweet edge dividing you through the heart and marrow.
—Henry David Thoreau, *Walden*

A number of years ago I was talking to a friend about a philosophical problem. I emphasized how the issue had perplexed thinkers throughout the ages, and how the accumulation of thought over the centuries had only rendered the problem that much thornier. I urged her to appreciate the depth of the dilemma and take up the challenge of resolving it. Not ignoring my sense of purpose, she nevertheless turned to me at one point and said, "Paul, this is where you and I part company. At this stage in the discussion you look at the unresolvable nature of the problem and say, 'Oh no!' I look at it and say, 'Oh well.'"

Many of us worry about things that others consider irrelevant—both philosophical and empirical. Some of us get wrapped up in abstract questions that others find abstruse, and many of us are disturbed by social dilemmas to which our friends and colleagues are indifferent. The combination describes many an environmentalist. Environmentalists worry about the well-being of the earth. We care about fresh air, clean water, healthy soils, and the planet's capacity to support human life and ecological abundance. While these are genuine concerns, there is something nonetheless abstract about them. Few of us live on the front lines of severe environmental harm, and our concerns for the earth as a whole are fundamentally theoretical insofar as no one can see and experience the planet in its entirety. Yes, we have viewed those famous photographs from space, and have seen pictures of the Katrinas, Chernobyls, and clear-cuts of the world. But to capture these in thought and develop a sense of care about them requires us to rise above our immediate experience. When we do so, sadly not everyone joins us. Distressing about the earth is often a lonely exercise in abstract worry. Yet worry we do. We see lots of things happening that don't sit right with us.

The great environmental writer, Aldo Leopold, starts his classic *A Sand County Almanac* by saying, "There are some who can live without wild things, and some who cannot. These essays are the delights and dilemmas of one who cannot."[1] Leopold was keenly aware that not everyone shared his love and sense of concern for the natural world. He wrote that "one of the penalties of an ecological education is that one lives alone in a world of wounds."[2] Loneliness seems to be a characteristic of the environmentalist life. Environmentalists frequently see themselves as modern-day Cassandras—the ancient Greek mythical figure who was given the power to foretell the future, but later punished with the curse that no one would believe her. While societies throughout the world have certainly heard the calls of environmentalists, they have yet to embrace environmentalism's sense of urgency and the depths of its commitment. The machine—the economic, cultural, and political system that inspires people to do, get, and yearn for more, with often disastrous consequences for both humans and nonhumans alike— rolls on. "Oh well"?

This book shares much with Leopold's sensibility. It is a book of worries about wild things and quite a few abstractions. Like many other American environmentalists, I value wildness. I love those aspects of life that we meet at the edge of our ability to understand and control. The nineteenth-century philosopher and naturalist Thoreau famously wrote, "In wildness is the preservation of the world."[3] Lots of people have tried to figure out exactly what he meant. To many environmentalists it is pretty clear. Humans thrive when in contact with the feral, spontaneous, or unbidden aspects of the world; we find much pleasure, greater sensitivity, and deeper levels of experience when we encounter things that elude our conceptual grasp and empirical control. For Thoreau and others this happens most palpably in the natural world. Mountains, waterfalls, elephants, and orchids seem to operate independent of human will. They appear to have their own way about them that makes them difficult to wholly size up. Preserving this wildness or otherness was important to Thoreau, and has been essential to the American environmentalist tradition. Because of nature's sheer otherness, Thoreau looked to it as a place to escape from the innervating pressures and entertainments of human society. It has long represented a realm uncontaminated by the pettiness that can often characterize human relations and a place where one could, as Thoreau puts it, "live deep and suck out all the marrow of life."[4] Wildness is the preservation of the world because nature's apparent difference—its nonhuman quality—opens us to a world different from and broader than ourselves. It thus prods us to marvel at existence itself and experience the sense of surprise and authenticity such wonderment frequently invites.

For many American environmentalists wildness also preserves the world in a more prosaic sense. As biological creatures, we humans need to eat, drink, and breathe to stay alive. Our ability to do so and with a decent quality of life depends on nature's regenerative, other-than-human capacity. Yes, to stay alive we often need to battle nature—to protect ourselves from the elements, wild animals, and vagaries of a world indifferent to our survival and well-being. But we must never forget that ultimately, at least at the biophysical level, we are subject to the earth. It provides natural resources, absorbs waste, and maintains a host of ecosystem services that are prerequisites to human survival. In this sense, the wildness of the earth is not so much a turn-on as a foundational ground on which we rely every minute of our lives. The wildness of nature is essential to our welfare and sheer survival.

The End of Nature

The premise of this book is that the wildness of nature, so dear to American environmentalism, is coming undone. Over the past century or so the human world has encroached on and ultimately colonized nonhuman life to the point where we are increasingly being denied access to the untamed or unbidden. The dream of modernity, in which, as seventeenth-century philosopher René Descartes notes, we "render ourselves, as it were, masters and possessors of nature," is quickly becoming all too true.[5] We are not only controlling nature but also wholly transforming it, and this makes identifying and securing its wildness that much more difficult, if not, for some, impossible. To use an all-too familiar environmentalist term, we are literally and figuratively *consuming* the wildness of nature. Our minds are taming it; our technologies are rendering it usable; our affluence is exploiting it; our power in general is transforming it. This book contemplates and worries about what that means.

Proclaiming that the natural world is disappearing and that humans will be worse off for it is nothing new. As far back as Plato people have complained about humans altering nature beyond repair or overlaying human thought too thickly on the nonhuman world so as to rid ourselves access to a realm independent from human society. To some, this has meant the disenchantment of the world; to others, it has meant a pressing threat to human physical well-being and survival. In either case, worries about disappearing wildness and the human-induced transformation of nature have been long-standing. They have been especially central to a prominent wing of the American environmental movement.

Today, however, those worries take on a new sense of urgency, and we need to develop a more philosophical appreciation for what is at stake. These days it appears that we are on the edge of not simply attenuating wildness from our experience but altering it beyond recognition. We are so thoroughly decimating the empirical reality of nature and so radically revamping our ideas of it that the whole ensemble of nature as that which is separate from humans is apparently vanishing before our eyes. If we are not mindful of this and vigilant in somehow protecting the wildness that is most easily accessible in nature, we will soon be writing an obituary—that is, if we are still around, able to hold a pen, and capable of imagining a world separate from ourselves.

Today's threats to the wildness of nature are taking place on two fronts: the empirical and conceptual. Empirically, a growing human population, unparalleled technological prowess, increasing economic might, and an insatiable consumptive desire are propelling us to reach further across, dig deeper into, and more intensely exploit the earth's resources, sinks, and ecosystem services. To be sure, humans have always altered nature. It is one of the paradoxes of life that we always change the very world on which we depend; simply being alive requires us to alter the natural environment. Recently, though, the cumulative force of our numbers, power, and technological mastery has swept humans across and deeply into all ecosystems to the point where one can no longer easily draw a clean distinction between the human and nonhuman realms. Whether one looks at urban sprawl, deforestation, loss of biological diversity, or ocean pollution, it is clear that humans have been progressively overtaking large swaths of nature and thereby imprinting themselves everywhere.

The empirical diminution of nature was given its most popular and forceful expression years ago when environmental writer and activist Bill McKibben originally published his book *The End of Nature*. McKibben announced that humans have exerted so much influence on the planet in recent decades that nature is not only shrinking but being wholly colonized by human beings as well. We mine the earth's crust, fish its oceans, pollute its air, reroute its rivers, and rework the land with such intensity and extensiveness that there are essentially no pristine landscapes, untouched wilderness areas, unfished seas, or even unobstructed skies left anymore. (We need only look outside to confirm McKibben's thesis.)[6]

McKibben claims that the ultimate death knell of nature is anthropogenic climate change. The buildup of carbon dioxide and other greenhouse gases has altered the temperature, humidity, and weather across the globe so much that every region and living creature on earth has been altered, if only ever so slightly (for now), by human activity. According to McKibben, climate change has definitively erased the distinction between the wildness of nature and the "made-ness" of human enterprise. We can no longer wake up in the morning and comment on what a beautiful day the earth has given us, or how special certain plants and animals are. We must now acknowledge that we have partially manufactured the natural world. For McKibben and others, the scope and scale of human activity has created a world in which there is no longer any such thing as nature devoid of human influence. Wildness, as that dimension of nature that signifies genuine otherness, has been stamped out now that the human signature can be found everywhere.

As if the physical disappearance of nature is not enough, certain intellectual understandings are declaring the conceptual end of nature. Most of us, including McKibben, are accustomed to thinking of nature as an independent realm that operates according to certain principles and possesses a given character. The whole notion of wildness is in fact premised on this orientation insofar as it suggests that nature has a particular way about it that is separate from and indifferent to human beings. Nature, in this sense, is the world acting by itself; it is the way of things beyond the human realm. These days, many prominent thinkers are pointing out that this view is anachronistic, if not fundamentally naive. Nature is not a self-subsisting entity with an essential character but rather a contextualized *idea* through which we approach the nonhuman world. That is, nature is not something laid out before us that we can apprehend in an unmediated manner; it is instead a projection of cultural understandings specific to certain times and places. In other words, nature is a social construction that assumes various meanings in different contexts. Thus, while Thoreau may have seen wilderness as a place of refuge, centuries before him others saw it as a foreboding area full of dangers in which one could literally and figuratively get lost. Likewise, today one person's endangered species is another's source of income, and what some take for a forest habitat others see as timber and board feet. Nature is not simply a material substratum whose essential character we glean from study and observation; rather it is a repository of meaning. This line of thinking rejects the idea that nature—as valued wildlife, beautiful landscapes, or simply a realm that holds specialness for humans—is disappearing, since nature *as such* never existed in the first place.[7]

Nature as social construct was given its most articulate and widespread expression a number of years ago in a set of essays, edited by historian William Cronon, titled *Uncommon Ground: Rethinking the Human Place in Nature*.[8] Cronon's volume presents an array of voices, coming mainly out of the humanities, that explain and illustrate the notion that "nature" is, fundamentally, an idea. The essays point out the many ways that nature as a locution is used to understand, describe, and appreciate the nonhuman world. This orientation doesn't mean, of course, that nature is a figment of our imagination or somehow does not exist. For example, the authors within Cronon's book recognize that Yosemite is a real place. They demonstrate, however, that Yosemite has come to mean particular things to those who visit, read about, or otherwise come to know it, and it is the contingency of such meanings that reveals nature's socially constructed character. As Cronon puts it, "Yosemite is a real place in nature—but its venerated status as a sacred landscape and national symbol is very much a human invention."[9]

The "end of nature," in an empirical sense, and what can be called social-constructivist "ecocriticism" fundamentally challenge our notions of wildness, and as a consequence, the foundation of American environmentalism. Whatever else American environmentalism is—and it is a good many things—at its core it is about nature. Since the environmental movement's early days in the nineteenth century when people began worrying about rural areas being colonized by encroaching industrialization, to contemporary efforts at addressing climate change, water scarcity, and ozone depletion, a significant strand has seen human well-being and survival wrapped up with protecting the nonhuman world. Sometimes, especially recently, this has expressed itself as a matter of concern for the way people use nature as an instrument to exploit others, or for how racism, poverty, war, and human rights abuses are connected to the control over, access to, or simply diverse experiences of land, water, air, species, or resources in general. Nonetheless, threaded through all of these is a focus on the way the material earth is used, altered, owned, accessed, preserved, or degraded. While the movement has thus changed over the years and today exhibits tremendous diversity, nature—and especially the wildness or otherness of nature—still stands at its center. Nonhuman nature provides the raison d'être of much American environmentalism. Given this, the end of nature raises important questions about the identity and future of the movement.

Without nature, what *is* the American environmental movement? On what philosophical grounds can it base its insights and construct its political strategies? Need it close up shop and go home since it no longer enjoys a reference point, or is there a new horizon toward which it should tack? Likewise, should we walk smilingly into a postnature environmentalist future in which we accept the dewilding of the world and simply make do, or should the movement continue to resist human encroachment on nature even though doing so these days is seemingly a quixotic task?

This book wrestles with these questions. It seeks to understand the significance of contemporary criticisms of nature, and reflect on what they mean for the future of American environmentalism. Its central argument is that the end of nature, while fundamentally challenging to the movement, represents not a death knell but rather an opportunity. It offers the chance for the movement to think afresh about conventional philosophical and political categories, and therewith refashion itself into a more effective movement. The end of nature, in other words, far from representing the demise of environmentalism, embodies the movement's future. As will become clear as the following chapters unfold, coming to terms with wildness in a world seemingly determined to snuff wildness out of our lives represents the promise of environmentalism.

Into the Postnature World

Simply to raise questions about the end of nature and fate of wildness at this point in time makes many environmentalists uncomfortable. Environmentalism in the United States is finally finding its political footing after nearly a decade at the margins of political life. The Bush administration started the new millennium with essentially a frontal attack that put environmentalists on the defensive as the White House aggressively sought to dismantle years of legislative environmental protections and international commitments.[10] The Obama administration has largely reversed this effort, as it has tried to usher in a new era of environmental responsibility. It has, for instance, increased the number of wilderness areas, tightened regulation over various pollutants, advanced a new energy policy aimed at reducing our dependence on fossil fuels, and tried to assume a leadership role in international environmental issues. To be sure, many of its efforts have been stymied, and environmentalists who helped put Obama in office expect more from him. Nonetheless, on the whole, American environmentalism seems to have finally found a hearing in Washington. Indeed, not since the 1970s have the political stars been aligned to enable environmentalism's message to be heard and its recommendations adopted. Given this, the last thing the movement seems to need right now is to get lost contemplating what seem like philosophical problems. Having at long last come out of the political desert, it seems unwise to rethink the movement's identity and political orientation. Doesn't the movement have enough on its plate right now advancing initiatives with a sympathetic administration?

A related set of concerns emerges as one recognizes the split character of the American environmental movement. While largely supportive of each other, many activist groups and intellectuals within the movement have made a profession of criticizing their brethren, and despite the new political promise of environmentalism (or because of it), such criticism continues. For instance, grassroots activists criticize national and transnational organizations for being out of touch with the needs and political visions of people on the ground. They complain that the sheer size of such organizations—often called "Big International Nongovernmental Organizations"—and the overly narrow agenda that such groups pursue alienate many would-be supporters, thereby compromising the movement's ability to mobilize people for significant campaigns.

From a different corner, radical parts of the movement criticize their more moderate counterparts for pursuing only modest goals in the political arena and nuzzling up too closely to industry in the economic sphere. They reject the kind of market-driven approach adopted by many groups that work with the business community and are wary about environmentalism's too cozy relationship with the Obama administration.

By way of response, the moderate or light green groups look at their radical counterparts as out of touch with political reality. By calling for global transformation and at times undertaking extreme tactics to make themselves heard, the radicals, according to their critics, render themselves politically irrelevant at best and invite backlash at worst. This is especially troubling when there is a sympathetic administration in the White House.

It is in this context that many environmentalists fear recent criticisms of nature or musings about the end of wildness. There may in fact be times to reflect on environmentalism's fundamentals and even rethink its raison d'être. That time, however, is not now. Rather, this is the moment to reengage and deepen long-standing efforts to protect wildlife, fight against pollution, safeguard natural resources, and support sustainable development. For many in the movement, to get wrapped up in an abstract debate about the status of nature represents a detour that the movement doesn't need right now and from which it may never return.

I see things differently. Environmentalism certainly has a unique opportunity to intensify its conventional efforts and move beyond the holding pattern it was experiencing until recently. Moreover, its internal squabbles go way back and thus do not pose significant, timely challenges. But this doesn't mean that all is rosy. Despite recent legislative and executive victories, and the mainstreaming of an environmental sensibility across the United States, environmentalism's prospects are still rather dim. Environmental issues continue to be overshadowed by concerns about terrorism, the economy, conflict in the Middle East, and other so-called high politics issues. Additionally, Obama's policy agenda is so ambitious that his environmental commitments are always being balanced and often compromised by other concerns. Combine this with the monumental scale of environmental problems and it is clear that we are not unambiguously on a steady road to a green world. The movement may be having its moment, but behind this the machine is still very much rolling on.

To me, this situation calls out not for circling the wagons and simply reasserting past strategies but instead for thinking afresh about the movement's core principles and therewith exploring new terrain. It is in this context that the end of nature, far from undermining American environmentalism, represents a profound opportunity. For far too long American environmentalism has placed nature on a pedestal and relied on it to advance environmental protection. Nature has been the standard against which to measure environmental degradation, the good toward which environmental policies should aim, and the realm most deserving of protection. While such a focus has achieved much, it has also restricted the movement's political reach and effectiveness by helping to polarize political debate—pitting the well-being of nature against that of humanity—and leaving the movement with a unidimensional philosophy that unnecessarily offends movement critics. More generally, it has imprisoned the movement in a certain historical era and conceptual framework such that environmentalism's voice, while certainly part of contemporary political discussions, is sounding increasingly anachronistic and actually less responsive to the growing enormity and complexities of our environmental challenges.

The end of nature offers—indeed demands—a new orientation. It presents the chance for the movement to liberate itself philosophically and politically from a nature-centric perspective, and thus cultivate frames of reference as well as devise strategies for creating ecological and social health in a world where it is impossible to separate humans and nature. To the degree that environmentalists recognize the hybrid character of human-nature relations and appreciate the end of nature arguments, they can self-consciously work to protect the well-being of *both* people and the nonhuman world, and capitalize on environmental protection opportunities that arise at the complicated interface between the two. As I will show, parts of the movement are already embarking on such a postnature environmentalist trajectory. One sees this in environmentalist efforts to address urban sustainability, social justice, poverty alleviation, and the rights of indigenous people. An understanding of the end of nature can further such engagements and reinvigorate the movement. It can encourage American environmentalism to get to know itself again in a changed political, biophysical, and conceptual landscape, thereby resetting its political compass. This can lead to a renegotiation of the fault lines that have long animated environmental politics and enable the movement to reposition itself so it can be more relevant to contemporary struggles. In short, rather than become nervous, environmentalists should embrace the end of nature, and take advantage of the opportunity it offers to become philosophically clearer and politically more effective.

What Do We Make of the End of Nature?

Such an embrace, while promising, will not be easy. Aside from pragmatic considerations, many thinkers and activists have responded coolly to the end of nature arguments on more abstract grounds. This is because the critiques cast into doubt not simply environmental thinking but also our ideas more generally about the place of humans in the world. People thus have been responding to the challenges posed by the end of nature in importantly different ways. For example, when it comes to concerns about pushing nature to the edges of the planet and, in the extreme, fundamentally stamping it out of our lives, some choose to put blinders on. Yes, humanity is trampling on wildness as people spread themselves across all parts of the planet and, yes, this entails a blending of human artifice and natural places. Nature is still very much around, however, and different enough from humans to relieve us of getting overly concerned about the end-of-nature arguments. There are still plenty of beautiful places and nonhuman species about. These may be flecked with a human influence, but the impact is usually so slight that dwelling on it is beside the point. Yes, climate change, for instance, has resulted in greater glacial melt, but this hasn't fundamentally changed the direction or flow of the planet's waterways, and thus it should not alter the way we understand glaciers, rivers, watersheds, and the like. The world has not become an artificial entity; the otherness of nature is still very present. Those subscribing to this view see humans and nature as fundamentally different things and understand that, while the two will occasionally mix, this doesn't render all parts of the earth human. In most instances, the mixing can't be detected and often can eventually be undone. This view doesn't completely dismiss the empirical end of nature but simply sees such concern as wrongly oriented. Rather than give up on nature, we should work to protect those elements that have yet to be significantly altered by humanity.[11]

Other people take a different tack. For example, many shrug off lamentations about disappearing nature out of the belief that humans are themselves natural, and hence cannot be faulted for extending the range of their presence or otherwise intervening in nature. In fact, the idea of intervening itself is meaningless. Humans are biological creatures just like all other organisms, and as such it is silly to say that they are encroaching on a world of which they are a part. Those advancing this view understand nature as everything that exists. "Nature means the sum of all phenomena, together with the causes which produce them; including not only all that happens, but all that is capable of happening," writes nineteenth-century philosopher and political economist John Stuart Mill.[12] Using this understanding, it is clear that humans cannot possibly alter the natural world since we are part and parcel of it. There is therefore no reason to worry about the end of nature because whatever is overtaking nature is indeed nature itself. Human colonization of "nature"—in the form of overfishing, suburban sprawl, air pollution, and so forth—is merely another evolutionary wrinkle in an ongoing story of ecosystem change. It's no big deal.[13]

Still others think that the end of nature *is* a big deal, but welcome it as an advancement in human well-being. Most people like the human-made world. They enjoy the security and comfort of living in an apartment or home with a reliable source of food and protection against the elements, and pine for these things when denied them. Most of us appreciate that the privileged among us are no longer subject to, say, the darkness of night (we can turn on lights), the culinary constraints of the seasons (we can import food from around the world), or the mercy of the weather (we have furnaces and air conditioners). These enjoyments have come through our battles with nature, and rest on such victories. Nature is, in so many ways, a constraint on our lives. Taking it over has thus been one of the most liberating human achievements. When McKibben and others complain about the end of wildness, they should take a hike to places where humanity has yet to master the natural world. Then they would realize that the end of nature is an event to celebrate, not lament.[14]

A final response sees the disappearance of nature as an important achievement, but is less sanguine about the gifts it will bestow on humans. Rather, many view the end of nature as an inevitable result of age-old human intervention into nature, and contend that whether we like it or not, we must now rise to the level of responsibility that taking over nature entails. They remain mixed, however, about whether we'll be able to assume this responsibility. For millennia, people have changed the earth. As mentioned, it is part of life to alter the world in which we live. Native peoples used fire to clear and fertilize the land, colonizers transported diseases and invasive species that reshaped ecosystems, and agriculturalists altered the evolutionary track of the plant and animal worlds through selective breeding. That we have altered the entire earth is thus no surprise, yet neither is it something simply to celebrate. As environmental scientist Daniel Botkin writes, "Nature in the twenty-first century will be a nature that we make; the question is the degree to which this molding will be intentional or unintentional, desirable or undesirable."[15] Or as botanist Peter Raven says, "We human beings are in fact managing the entire planet Earth, every square centimeter, right now, and the illusion that we are not, that any one of us can be exempt from this work, is extremely dangerous."[16] In other words, the end of nature changes our historical role on earth to the degree that it calls on us to consciously take hold of the steering wheel of life, and become intelligent, compassionate, and otherwise mindful managers of the planet—quite a daunting challenge.

These different views suggest that coming to terms with the end of nature—either believing that it has actually happened (or even could happen) or pragmatically responding to it—is no easy matter. While some reject the whole idea that nature can be completely colonized, others embrace it as reality (and for some, a desirable reality). Such disagreement is not surprising given the stakes involved. The end of nature argument is not simply an environmentalist worry. It is also about the fundamental meaning of human life on earth.

The same uneasiness or diversity of opinion exists with regard to social constructivist ecocriticism. Constructivists claim not that nature is empirically disappearing but rather that it never really exists separate from the interpretative meanings we give it. This has elicited two main responses. On the one hand, many dismiss the social constructivist attack on nature out of hand, and merely reassert a modernist narrative about nature and its imperatives. Many see ecocriticism as a type of environmental relativism that is at odds with common sense and contemporary science. Certainly, they acknowledge, there is a social dimension to how we think about nature, but nature is fundamentally a physical entity, and our understanding of it can be based on scientific description. The whole notion that nature is constructed is intellectual sophistry practiced by ivory tower geeks who never venture outdoors or work at such high levels of abstraction that they never genuinely engage the phenomenal world.[17]

Others see attacks on the idea of nature as simply the latest manifestation of a long anti-nature tradition associated with what biologist David Ehrenfeld calls the "arrogance of humanism."[18] Ecocriticism places human beings at the center of all phenomena and hence is overly impressed with the self-referential character of human experience. Consequently, it is blind to what philosopher Albert Borgmann calls nature's nonhuman, "commanding presence."[19] Those who feel this way argue that ecocritics are wrong in their so-called insights. Ecocritics practice "fashionable nonsense" as they overemphasize the social dimensions of the scientific enterprise and, out of a desire to appear intellectually cool, join the academic chic in getting rid of the nature of nature.[20] As such, many claim, we must reject (and therefore not worry about) their proclamations.

On the other hand, many go in the opposite direction and fully embrace the constructivist critique of nature. They maintain that since everything we call nature is relative to our ideas, we should accept and even celebrate our role as its creators. Like those who support the empirical end of nature and urge humanity to assume a managerial position, those excited about ecocriticism recognize that humans have always altered the material conditions on earth, but have done so within particular discursive contexts. We can never escape these sociohistorical cognitive landscapes and therefore should not try to do so. Instead, we should accept the constructed character of nature, appreciate the ways we invest the nonhuman world with particular meanings, and get on with it. Getting on with it entails doing whatever we want—which usually means utilizing technology, contenting ourselves with human-made landscapes, and happily using artificial substitutes for natural resources. For nature is not some other-than-human world that we find but rather part of the world we make. We should, to be sure, make our world into a place that maintains ecological services, yet our vision and control over the environment need not be hindered by any preconceived vision of what is natural. Political scientist and futurist Walter Anderson represents this view when he recommends that we see ourselves for what we really are: ecoartists—designers and builders of the nonhuman world.[21] Those espousing this perspective call for dispensing with the category of nature altogether, and fashioning an environmentalism that uses other guidelines for vision and mobilization.

There are no easy answers to the problem of nature's status as something wild and fundamentally different from humans. This should not surprise us in that thinkers have been wrestling with the concept of nature for centuries. As social critic and novelist Raymond Williams writes, nature is "perhaps the most complex word in the English language."[22] One reason for this is that nature has so many meanings. Yes, it defines the world of plants, animals, mountain ranges, and so forth, but it also describes the "way of things," the patterns by which things "naturally" evolve or express themselves (including human beings). In *My First Summer in the Sierra*, naturalist John Muir observes, "When we try to pick out anything by itself, we find it hitched to everything else in the universe."[23] This describes the difficulty with coming to terms with critiques of nature. The end of nature argument and social constructivist ecocriticism do not stand in isolation from the long history of reflection on the nature of things or the commitments various people have made to the practice of nature protection. It would thus be astonishing if there was agreement about what to make of recent empirical and conceptual assaults on nature.

The difficulties involving nature go even deeper. It is not simply that various camps line up in opposition to each other but rather that the issues are so perplexing that individuals find themselves split. Thinking about nature and its end is so riven by paradoxes that it almost necessarily sends the mind spinning. If we are to avoid dogmatism, which as philosopher and dramatist Gotthold Lessing claims, identifies "the goal of our thinking with the point at which we have become tired of thinking," we must shy away from finding too easy a resting place in such issues.[24] As mentioned and as I will explain, recent criticisms of nature offer tremendous opportunities for both environmentalism and the struggle to find our place in the cosmos in our overly critical age precisely if we maintain a sense of ambiguity about nature.

Traversing the Human/Nature Divide

Like others, I too am torn by the critiques of nature. I resonate, on the one hand, with the above-mentioned criticisms. In my head anyway, I understand the arguments of those declaring an end to nature and revealing nature's socially constructed quality. Both perspectives are "right," in the sense that they are based on sound thinking, and informed by either careful observation or long traditions of philosophical thought. The wildness of nature *has* indeed largely disappeared as humans have placed their signature on all the earth's ecosystems. And we *never* come to nature unencumbered by cultural, personal, or subjective categories. To me, the ideas that nature is gone and that our conceptions of it are largely solipsistic are compelling. I also am sure that many environmentalists would concur with the logic behind these views.

At the same time, there is something inside me—and inside many American environmentalists—that finds both sets of insights unacceptable. A postnature world may make sense intellectually, but emotionally, morally, and I dare say spiritually it makes no sense at all. Emotionally, it offers a lonelier world in which we cut ourselves off from other creatures; morally, it appears overly anthropocentric and thus ethically arrogant; and spiritually, it makes us not simply an exceptional being in the universe but the be all and end all of existence. Many environmentalists have similar misgivings. In fact, a postnature image of the world is at odds with the central tradition of American environmentalism. A postnature world cannot possibly sit well with those who associate their lives with Thoreau, Muir, Leopold, Rachel Carson, and others who prize the other-than-human world as something biologically, morally, and spiritually essential to human life.

Because of the complexities involved with thinking about nature, my sense of being torn extends to the wider uncertainty regarding humans and nature. Like many American environmentalists, I enjoy both the experience of being in nature—camping, walking, or just being surrounded by the natural world—as well as the comforts of the human-made world. After a long backpacking or cross-country skiing trip, I love coming home to the warmth of a furnace, the ease of retrieving food from a refrigerator, the entertainment of a stereo, and the shelter of walls decorated with various pieces of art. Thoreau celebrates this dual aspect of experience in his essay "Walking." On returning from a multihour walk through the woods, Thoreau remarks how much he enjoys his abode, where he can settle in front of the fire, escape the wind and rain, and take pleasure in reading or writing free from the elements.[25] Indeed, most of us, no matter how long we leave society for the wilderness, enjoy coming back. Few of us are willing to live in the wild all the time.

For many of us it is not simply a matter of returning to human-made comforts. These days, we take such comforts along with us. I go to the woods in stitched leather boots, a GORE-TEX jacket, and polypropylene underwear. I sleep on a synthetic pad, carry a metal canister of butane gas, and use a plastic water filter filled with chemicals to treat the "natural" condition of streams. If you look in my backpack, you'd find a whole host of items—from toilet paper, plastic bags, and rain gear, to flashlight, freeze-dried food, and plastic scouring pad—that are human-made, but that make being in the woods that much more enjoyable. Am I being hypocritical in my wilderness experience? Frankly, I'm unsure how one could be otherwise. Being torn about the criticisms of nature, in other words, reveals tensions about our relationship to nature more generally.

I was taught that when faced with contradictory feelings about something, or when mindful of paradoxes and tensions in the world, it is best not to resolve them too quickly less we forego the opportunity to learn something about the incongruities themselves. This is especially the case with regard to the question of nature and a possible postnature world. The struggle between the head and heart in this context offers a chance to think and feel afresh about environmentalism as well as the human condition more broadly. It prods us to resist analytic rigidity in which we must draw a sacrosanct line between humans and nature, or else let such a line completely disappear. What happens if we do neither? As I hope to show, such mindful resistance opens new vistas for environmentalism—vistas that offer greater conceptual clarity for a movement and world poised at the edge of a postnature age.

The Dual Dreams of Naturalism and Mastery

Resisting the impulse to resolve dilemmas is not easy. Few of us are comfortable with ambiguity, and nuance seems only to weaken political intent. It should be no surprise, then, that despite the tensions just described, in debates about nature people usually come down on one side or the other. We see this in the fault lines that have come to define American environmental politics.

Environmental politics in the United States are exceedingly polarized. Environmentalists are labeled "tree huggers," and accused of caring more for plants and animals than people; skeptics, on the other hand, are seen as greedy technophiles blinded by self-interest. Such polarization is not simply caricature but instead reflects genuine disagreement. Environmentalists and their critics argue about a lot of things. They disagree about land-use issues, climate change dangers, the value of biological diversity, the use of toxics, and the costs of mountain-top-removal mining. At bottom, however, such disputes are merely the circumstantial reflection of a deeper ideological clash having to do with two fundamentally different worldviews. These two views stand as oppositional narratives about the place of humans on earth. They offer ideological comfort for negotiating one's way through the perennial challenge of making sense of nature and humanity's relationship to it. The great promise of the end of nature argument and social constructivist ecocriticism is that they can help relax the rigidity with which these views are held. This would help the environmental movement by encouraging it to shed some of the baggage it has accumulated over the years, and therewith assist in resetting the fault lines that distinguish environmentalists from their critics and polarize political debate.

If you scratch American environmentalists deep enough, you'll find a basic sensibility that informs many of their understandings and practices. For reasons to be explained, I call this sensibility the "dream of naturalism." The dream of naturalism believes that the best thing human beings can do is to align themselves with the imperatives and patterns of nature. It assumes that there is a world outside of human beings, and that this world sets the parameters and provides special promise for human life. It recognizes that the more-than-human world has perfected, over billions of years, ways of creating and sustaining life, and that we should respect, appreciate, learn from, and emulate its workings rather than try to outsmart them. The reasoning is pretty basic. Nature provides the biophysical requirements for human life, and we ignore its imperatives at our peril. We need fresh water, healthy food, clean air, and the like to live and thrive. If we undermine these, we suffer and, in the extreme, endanger our very survival. The reasoning gets deeper, though, as we realize that nature means more to many of us than simply a biophysical backdrop for our lives. Many of us in the American environmentalist tradition look to nature as a model for living, ground for morality, and source of aesthetic pleasure. Nature, in other words, is not simply a material substratum that we live within and depend on but also a source of principles, cultural edification, and delight. It represents much that is true, good, right, and beautiful in the world. As a consequence, we should harmonize ourselves with, rather than impose ourselves on, the natural world.

Environmentalism's embrace of naturalism leads it politically to support policies that protect nature's otherness. Environmentalists prefer the earth's air, water, soil, and species as "given"— uncontaminated or at least not too altered by humans. We rail against pollution, anthropogenic climate change, and the loss of biological diversity at the hands of humanity. Indeed, most environmentalist campaigns, particularly in the United States, have an element of protecting nature unto itself. Nature not artifice should be our guide in environmental matters.

Critics of environmentalism subscribe to a radically different orientation. Far from being a sacrosanct realm that deserves pragmatic deference and principled consideration, critics see nature as merely the biophysical backdrop for human life. There is nothing particularly special about it in a philosophical sense; if anything, the natural world is merely raw material ripe to be used and designed as humans see fit—that is, it is there for the taking. To the degree that nature has any character at all, it is as constraint to be overcome. Wind, rain, wild animals, excessive cold, and so forth curtail human freedom, and even threaten our survival. Human well-being consists of freeing ourselves from nature's exigencies—opting out, as much as possible, from nature's imperatives. Critics of environmentalism resonate with Francis Bacon's dictum that nature should be "bound into service" and "made a slave."[26] As such, they subscribe to what could be called the "dream of mastery."

The dream of mastery turns on the notion that human beings are uniquely endowed with ingenuity, resourcefulness, and the spirit of enterprise, and that we can and should use these to unlock and override nature's secrets. When we do so, we improve human life. Medical technologies, agricultural sciences, electronics, and the like are all examples of humans bursting through previously established biophysical limits, and each has brought humanity much safety, comfort, and delight. The dream of mastery seeks to deepen our commitment to humanity's quest to decipher and control nature. It expresses itself politically through policies that unleash science and technology, and cultivate a spirit of human confidence and enterprise. It sees humanity, as business economist Julian Simon puts it, as the "ultimate resource," and thus able to address any challenges, including environmental ones, simply through the application of greater human effort and control.[27]

The twin dreams of naturalism and mastery are, of course, ideal types.[28] They represent broad interpretative strokes trying to depict philosophical proclivities. Nonetheless, they capture the dual sensibilities that inform and animate American environmental politics. Environmental disputes are in many ways arguments about fundamentals. They involve clashing worldviews of humanity's place on earth. As such, they are almost theological in character. They pit the godly character of nature against the godlike attributes of human beings. Environmental politics in the United States has been mired for too long in an endless debate about which god is, or should be, primary. The end of nature and social constructivist ecocriticism can help advance such debate by stripping each camp of its theological assumptions. As I show in the following pages, such a contribution would enable American environmentalism to better understand and position itself in the postnature world, even if in doing so it must embrace a type of ambivalence when it comes to ultimate questions about our place in the more-than-human world.

Living through the End of Nature

Most of us hate ambiguity. We like feeling certain about ourselves and the world, and flourish to the degree we feel confident in our life paths. Living day-to-day involves negotiating our way through complexity, and often struggling against forces that threaten to alter our lives in basic and not always attractive ways. Ambiguity seems to undermine our abilities.

Environmentalism as a social movement is no different in its desire for certainty. Environmentalists engage complicated issues and battle an array of powerful forces. Many people are tone-deaf to environmental issues, or simply too wrapped up in their personal interests to care about the well-being of the earth's life-support systems, environmental injustice, or the prospects of future generations. Additionally, the trends of population growth, increasing affluence, and technological wizardry are intensifying almost all environmental challenges, forcing environmentalists to concentrate on a moving target. There are also structures of power—associated with capitalism, the nation-state system, patriarchy, and modernist, scientistic logic—that generate environmental degradation and demand a response, but defy easy analysis. In the face of such complexity and forces, there is the urge and seeming necessity to develop a keen-eyed sense of "what is to be done," and advance such thinking in a fierce and frequently uncompromising manner. As environmental activist David Brower once remarked, "Polite conservationists leave no mark save the scars upon the Earth that could have been prevented had they stood their ground."[28] Standing one's ground is not something we do well when we are ambivalent. Ambivalence seems to make us vulnerable to being swayed off our path. It involves uncertainty and doubt, and engenders hesitation and indecision. The last thing environmentalism seems to need these days is ambivalence.

Yet ambiguity may be the movement's saving grace. The world is a complex place. There are no easy answers to many of the issues environmentalists wrestle with—at multiple levels of concern. Is globalization good or bad for the environment? Should environmentalists depend on scientific logic to advance environmentalist concerns, or is scientific logic itself part of the problem? How should environmentalism engage capitalism? Should it work in tandem or seek to overthrow it? Can the international state system address global environmental issues, or should we seek a new world order with different types of political units? Is technology good or bad for environmental well-being? Negotiating our way through such questions cannot be a matter of ideological bulldozing but instead must involve nuance, contextualized thinking, openness, and at bottom, a type of faithful unknowing. Environmentalism is often scared to advertise its own uncertainty about issues as it has been fearful, more generally, of exposing rifts within the movement. An appreciation for the end of nature arguments suggests that this is a mistake.

The empirical end of nature and social constructivist ecocriticism offer us new ways to think about ambiguity, and especially ambiguity in a political context. There is no straightforward answer to the perplexities of nature. As we shall see, we cannot come down definitively on the question of whether we are part of or distinct from nature, or whether we should lord over or subject ourselves to nature as evidenced in the debate between naturalism and mastery. Likewise, there are no absolutes when it comes to thinking about the social construction of nature. Yes, we humans tell ourselves stories about the more-than-human world, but there also seems to be something genuinely revealing about those stories. They are made-up and seemingly true at the same time.

In the following pages, I highlight uncertainties of environmentalism. I argue that these uncertainties are, paradoxically, entryways into a deeper kind of knowledge—one that better understands our inner lives and outer experiences as environmentalists. Ambivalence is not some horrid sensibility that makes us weak-kneed and ineffective. Rather, it is a source of wisdom and, as I will assert, political strength. Life is full of mysteries. We may know that we evolved along with other creatures, and that our bodies operate according to physical and chemical laws, but we have no clue about what it all means, what is absolutely best for our lives, and how to pursue meaningful agendas in a world that is quickly changing, and in which we ourselves are shifting our affiliations as well as finding new passions and interests.

Many traditions have long maintained that the one thing we know about life is that things change. This is the one constant. Whether it's Heraclitus, the Buddha, or contemporary physicists, we know that circumstances are always shifting. Living in such a world is an exercise in openness and requires a confidence in unknowing in an absolute sense. Philosopher Alan Watts talks about the "wisdom in insecurity" to capture this.[29] The environmental movement has long appreciated ecological insecurity. These days, it is awakening also to movement insecurity. Environmentalism is increasingly uncertain about its core identity, which for centuries, at least in the American context, has revolved around the idea of nature. In working through this identity crisis, it cannot simply abandon the term and the reference toward which the word points, nor can it easily continue uncritically to embrace the idea of nature. We need to find a middle path. As I hope to show, this middle path is not mere "polite environmentalism" or a mishmash of muddled thinking. Rather, it is involves operating across the fault lines of philosophical contestation, and fashioning the tension itself into insight and practice.

The middle path involves living through the tensions of the end of nature. Those who wish to sweep the end of nature argument and social constructivist ecocriticism under the rug want to pretend that we still live in the twentieth century or even the late nineteenth one—a time when we could entertain a naive notion of nature and work productively to keep humanity out of areas long devoid of human presence. Alternatively, those who wish to embrace such arguments and go full steam ahead toward a postnature world are too willing to abandon values and understandings that have long inspired and informed humanity. The middle path is about holding on to both sets of sensibilities. It involves appreciating the contradictions that inflect the contemporary world—for example, protecting wildness by intensively managing wilderness areas—and those that mark our inner lives—for instance, loving both the experience of hiking through mountains and vegging out in front of the tube. Living the tensions of the end of nature calls on us not to choose sides—within ourselves or the external world—but to enlarge ourselves to include both sensibilities. A meaningful, effective environmentalism for the twenty-first century requires us, in other words, to maintain the intensity that contradiction provides, and milk it for insight and effective policy.

This book explicates the tensions of being an environmentalist in a postnature age. Its aim is to hold a mirror to ourselves, as environmentalists, so we can better understand ourselves within the complicated world we live in. The hope is that such an exercise will enable us to live more meaningful lives and invest ourselves in environmental protection in more effective ways. We are often told that the most useful kinds of books are those that simplify the world, those that reduce complexity so we can perceive the broad outline of things and thus understand life with greater clarity. This book takes a different tack. As I see it, many contemporary difficulties stem from pretending that life is fairly simple, that we can confidently understand its fundamental themes, and that we can therefore exert control over ourselves and much of our world in a self-assured manner. A necessary antidote to this is to complicate the world. This doesn't mean inundating the reader with more detail but rather offering ways to appreciate how intricate and ultimately mysterious life is, and how such an appreciation can enhance environmentalism.

Despite a deep attraction to the dream of naturalism, many of us, as environmentalists, live in two worlds: an ideal one, in which we respect, honor, and treasure nature; and a more pragmatic one, in which we constantly compromise our love of nature to get through our days. We care about other creatures and the earth as a whole, but we also like to get around on fossil-fueled cars and planes, eat exotic and nonlocal foods, and type books or simply surf the Net on computers rich in cadmium, lead, and barium. As I hope to show, these dual loves are not pathologies but rather genuine reflections of living in a post-nature world. We love the woods and our iPods. This doesn't mean that something is wrong with us; it instead expresses our environmental reality at this sociohistorical moment.

Likewise, environmentalism as a movement is split these days across two worlds. On the one hand, it wants to preserve, conserve, and sustain the more-than-human realm, which involves minimizing our presence, reducing our footprint, and otherwise restraining our interventions. On the other hand, we are realizing that this cannot be done without extreme intrusion using some of the most sophisticated technologies and managerial types of control. In the following pages, I show that these dual orientations are not antagonistic, even if they are on some level contradictory, but rather necessary practices in a postnature age. We live in a completely humanized world in which every corner of the globe has been inflected by human presence and in which our ideas have become so solipsistic that we can no longer see beyond our own social constructivism. In such a world, environmentalism can only operate stretched across constant tension. Anything else would be disingenuous. This book aims to articulate what it means, both individually and collectively, to live through the tensions of the present postnature age. It does so to help us deepen our experience of being environmentalists, and contribute to a more robust, historically relevant, and vibrant movement.

Sequence of the Argument

The book unfolds in the following way. In the next chapter, I provide a brief historical sketch of the environmental movement in the West, with special reference to the American context. I do so to highlight how the American environmental movement has drawn the distinction between humans and nature, and how this has served it in its political efforts. Since its early days, in the late nineteenth century through its contemporary expressions in the twenty-first century, American environmentalism has invoked a human-nature boundary to warn people against delving too deeply into the natural world. The boundary has been essential to cultivating a preservationist, conservationist, and sustainability ethic. The chapter aims to make this connection, and then explain the ways it has benefited the movement.

Chapter 3 explores what is behind the impulse toward such a boundary. Here is where I spell out the dream of naturalism. I explain that the boundary represents environmentalists' great love for nature—a love that borders on theological subscription. I describe various elements of this love: environmentalism's sense that nature is the true, good, right, and beautiful of the world. Appreciating the depth of environmentalism's adoration for nature is key to analyzing the challenges that the movement faces as we move toward a postnature world, and the distinct ways it can negotiate through the tensions that are increasingly becoming evident in a humanized world.

Obviously not everyone is a committed environmentalist, and thus to understand the political dynamics involved with moving toward a postnature world, one must appreciate other perspectives. The most important is that which environmental skeptics and critics of the movement espouse. This is where the dream of mastery comes in. In chapter 4, I point out that the dream of mastery, like its counterpart, relies on the boundary between humans and nature—only in this case instead of championing a policing of the boundary, it prescribes overriding it. As mentioned, in contrast to naturalism, mastery sees humanity rather than nature as the true, good, right, and beautiful in the world, and appreciating such privileging is essential for coming to terms with how the debate between the twin dreams of naturalism and mastery—a debate that fundamentally informs and animates environmental politics—has been playing out, and how the end of nature and social constructivist ecocriticism can shift the ground of such debate.

Chapter 5 begins to catalog such a shift. It explains how neither the dream of naturalism nor mastery can sustain itself in the face of contemporary events and ideas. It describes the ways in which humans are seemingly erasing the divide between humans and nature, and rethinking the fundamental category of nature itself. Here I detail what I have been calling the end of nature and ecocriticism. As mentioned, these critiques of nature threaten conventional environmentalism. They also offer the movement possibilities for refashioning itself to become philosophically more coherent and practically more relevant for contemporary times—ironically by embracing an ethic of ambiguity.

Chapter 6 continues to catalog the shift that the end of nature and social constructivist ecocriticism can help to instantiate. It does so by examining wilderness protection in a postnature moment. It describes conventional environmentalist orientations to wilderness and demonstrates how they no longer make sense. Informed by the dream of naturalism, much environmentalism has worked to cordon off remote or ecologically rich areas in the interest of preserving remnants of wild nature for human enjoyment and ecological health as well as out of a sense of moral obligation. These areas are guarded from human intrusion and theoretically preserved in their roughly natural state. In chapter 6, I look at the increasingly anachronistic quality of this approach. I explain how wilderness protection today involves a tremendous amount of human intervention. I make clear that wilderness as we know it is not left on its own but rather is highly managed using some of the most sophisticated forms of technology, capitalist models of resource use, and modernist sensibilities. Ironically, to preserve the wildness of wilderness these days, people have to engage in an awful lot of taming. The chapter does not stop there, however. It also points out that the management of wilderness, for all its technical skill and promise, can go only so far. Wilderness might be able to be managed, but it cannot be mastered in the sense of subjugating the nonhuman world to pure human design. Neither the dream of naturalism nor mastery is any longer appropriate for addressing questions of wilderness. The chapter ends with suggestions for crafting a postnature approach to preserving wilderness.

Chapter 7 provides a similar story with regard to climate change. Climate change is the most daunting environmental challenge. Much of life's future rests on how we approach it. Informed by the dream of naturalism, most environmentalists advocate getting out of the greenhouse gas business. Humanity should curtail and eventually halt our impact on the atmosphere. We should restrict ourselves from emitting too many greenhouse gases and let the atmosphere reconstitute itself. In contrast, subscribing to the dream of mastery, many suggest that we can continue using fossil fuels and even emitting greenhouse gases with the faith that we will simply technologically invent our way out of our troubles. Whether through geoengineering or some other technical feat, humanity will not have to alter its current trajectory. In chapter 7, I show that neither alternative holds much promise. The idea of leaving the atmosphere alone is no longer an option. Our interventions have brought us climate change, and no scenario even remotely being considered suggests that we can pull back enough to enable the atmosphere to "right itself." Moreover, there are significant questions to ask about what righting itself means: Are we aiming toward a preindustrial, prehistoric, or other state of affairs, and is *this* natural? Similarly, the dream of mastery is unpromising to the degree that it was the aspiration to mastery that brought us climate change in the first place. Applying simply more conquest as a way to rid ourselves of the adverse effects of conquest seems particularly troubling. The chapter ends with ideas about how to fashion a postnature orientation to climate change.

Chapter 8 concludes the volume. It articulates what I have been calling the middle path. This path is not an answer to our ecological woes or even a set of principles to inform environmentalist policies. Rather, it is a sensibility that one cultivates to live through the paradoxes of a postnature age. A postnature age is one in which neither nature nor humanity has a singular essence or fundamental nature. It is an epoch in which we are adrift from the theological categories that have long provided intellectual, emotional, and even spiritual insight and comfort. The middle path is an environmentalist trail through such post-theological terrain. Like all paths that lead to uncertain futures, it has no single map nor even a clear trajectory. One walks it, then, like all genuine paths, with mindfulness and heartfulness fully alive to the twists and turns along the way as well as the grit under one's feet. Such awareness does not bleach out past theological categories but instead removes such categories of their theistic authority, and thus opens our eyes more widely to the tensions that mark our world.

Physicist and energy guru Amory Lovins was once asked in a seminar, "What is the single most important thing an environmentalist can do today?" He responded with two words: "Pay Attention."[30] The middle path I describe in the last chapter is about paying attention in a postnature world. It involves maintaining a love for wild things and recognizing the impossibility of sustaining that love in a straightforward manner. Such is the challenge of any act of love. Such is the future of American environmentalism.

"The Bunker and the Sea," Cape May, New Jersey

2

American Environmentalism and Boundaries

But, man is everywhere a disturbing agent. Wherever he plants his foot, the harmonies of nature are turned to discords.

—George Perkins Marsh, *Man and Nature*

Environmentalism is among the most noble and generous of social movements. It cares about the biological foundations of life, works to protect both present and future generations from ecological harm and environmental injustice, and extends its sense of care to the nonhuman world. Translating this sensibility into a political campaign has not been easy. Politics eschews complexity. In the heat of competing interests, actors and movements must find ways of simplifying their message—a requirement that calls on them to conceptualize their concerns in broad strokes and package their public outreach in digestible ways. Like other social movements, environmentalism does this by drawing boundaries. It differentiates its agenda from others and articulates a distinctive view of things that is at odds with its opponents. Such distinctions, to be sure, while required of political participation, are not mere public relations; when expressed well, they reflect deep-seated insights. The most important distinction that American environmentalism draws is between humans and nature. By establishing a boundary between the two, American environmentalists are able to identify the stakes of their concerns and translate these into political expression. Nature provides the biological requisites for all of life and the medium through which much injustice is transmitted. Circumscribing it and distinguishing it from the human realm has been essential to American environmentalism's identity and political engagements.

The empirical end of nature and social constructivist ecocriticism cast doubt on the human-nature divide. As our species has extended itself across and into every ecological niche, and as we have come to understand nature as a realm onto which we project meaning, it is increasingly hard to maintain faith in and work politically with the boundary between the human and nonhuman worlds. The end of nature claim and ecocriticism both call on us to question the movement's commitment to the borderline. How crucial is the divide to American environmentalism's self-understanding and political efforts? Can the movement let go of it without risking philosophical coherence and political efficacy? If so, how much should it release its grip, and what gain can it expect? As I will show, the prospect of relinquishing the boundary, even a little, poses threats to the movement's identity and political effectiveness. Since its beginnings, it has enshrined the human-nature boundary as the central fulcrum of its politics. The divide has given expression to essential principles and guideposts of the movement—like preservation, conservation, and sustainability. Absent the divide, the movement will be at some loss in its political positioning and conceptual intelligibility.

Mindful of these issues, this chapter opens a conversation about the fate of the human-nature dichotomy in environmentalism's politics. It does so by identifying the formative events and thinking that crowned the divide as a central premise in the movement. While the end of nature and ecocriticism suggest that we rid ourselves of the boundary, it is worth recognizing the many benefits the divide has provided and the virtues it has made available, and how these have served the movement over its long history. An appreciation of these indicates that we should practice profound hesitation before letting go of the human-nature dichotomy even if contemporary affairs and thought deem it necessary.

Before explaining the evolution of the human-nature divide in American environmentalism, it is important to delineate the scope of this chapter and, by extension, the focus of the book more generally. Americans often think of themselves as living at the center of the world. American hegemony has extended its reach far and wide over the past century, and this creates the illusion that the rest of the world is merely playing cultural and institutional catch-up with the United States. U.S. hegemony also gives the false sense that American institutions and values are homegrown, arising out of the unique experience that is, we are told, "America." Such self-regard is frequently practiced when analyzing the American environmental movement. Many observers tell the story of American environmentalism as if the movement originated solely on American soil and possesses distinct U.S. ideals. To be sure, there is much that is unique about American environmentalism. In fact, as I will explain, the human-nature divide itself is relatively exclusive to the American environmental tradition. Nonetheless, the movement is not self-originating. It emerged out of a broader geographic context in which various environmentalist sensibilities have been expressed, activist campaigns have been waged, and practices have been developed. For this reason, while my concern is to explain how the boundary between humans and nature evolved within the American environmentalist tradition, it is necessary to cast the historical and conceptual net beyond U.S. shores. American environmentalism is a subset of environmentalism more generally.

While it is necessary to cast a wide net to appreciate the evolution of American environmentalism, it is also important to recognize the variegated character of American environmentalism itself. So far, I have been discussing the American environmental tradition largely as if it were a single entity, as if activists and thinkers through the decades and even centuries have shared a common orientation and agenda. Such characterization is obviously simplistic as it bleaches out significant differences within the movement. Not only has the movement always been rift with differences of outlook and political engagement, it continues to be so divided today. As mentioned, so-called light greens disagree with dark greens on everything from political strategy to philosophy. Grassroots groups take issue with national organizations, and urban-based activists often focus on radically different issues than their rural counterparts. Indeed, the environmentalist landscape in the United States is vociferously divided as the environmentalist agenda in general has broadened over the years to include issues of social justice, war and peace, corporate globalization, the rights of indigenous people, and sustainability in the broadest sense of the term. Given this, it is probably more accurate to talk about American environmentalist *these days than a singular movement. This is especially the case when analyzing the human-nature divide in American environmentalism. Plenty of formative thinkers and activists as well as many contemporary figures in the movement do not see themselves as protectors of the nonhuman world per se, and would take issue with viewing the movement as a whole as overly concerned with the fate of wildness.[1] Such voices are important, and I do not intend to silence them. It is nevertheless the case that a concern for the wildness or otherness of nature has long coursed through and continues to animate the predominant wing of the movement. As mentioned in the introduction, many American environmentalists see nature as their raison d'être, and certainly many observers understand American environmentalism as a nature movement. Thus, to ignore this element is to miss something critical about the evolving identity of American environmentalism. It would dismiss the challenges that the ends of nature pose to the movement and foreclose avenues of insight for thinking about the movement in a postnature age.

In the following, then, I both situate American environmentalism within a broad context and explain how the human-nature divide arose amid a complex, variegated movement.

The Boundaries of Early Environmentalism

It is difficult to say exactly when environmentalism first emerged. There have always been people concerned about what we now call environmental well-being, and hence the movement's lineage is both long and complex. The ancient Hebrews, for instance, created laws to prevent air and water pollution, and took measures to avoid destroying trees and nonhuman habitat in times of war. As well, Hesiod, of ancient Greece, expressed reverence for the earth, and recorded prohibitions against contaminating rivers and springs, and Plato wrote of severe deforestation, and lamented the loss of fertile soil and defaced landscapes. Indeed, a steady, if understated, discourse runs through history in which people have worried aloud about and taken actions to minimize environmental degradation.

Notwithstanding this, most observers locate the first inklings of the modern environmental movement in the rise of industrialization in the West. Starting in the late eighteenth and early nineteenth centuries, societies harnessed oil, gas, and coal in large quantities, replacing manual labor with steam-powered manufacturing, and expanding trade through a growing network of rail lines and roadways. Workplaces became mechanized, and trades were broken up into specified tasks. Together these allowed large accumulations of capital, which spurred further investment in manufacturing along with the emergence of financial markets through which goods and services could be bought and sold. Expanding markets drove demand and created consumer societies within which people had the means and manufacturers supplied the products to satisfy increasing material needs and desires. While industrialization certainly improved the lot of many—raising numerous people out of abject poverty and providing alternatives to agricultural work—it also had its dark sides, generating resentment and critique. Novelists such as Charles Dickens wrote about the inhuman working and living conditions in the cities of industrial Europe, and others like Mary Shelley, author of *Frankenstein*, portrayed the dangers of a machine world gone awry. Poets such as William Blake and William Wordsworth complained about how the noisy and dirty factories, mills, and railways of industrialization were degrading rural landscapes and agrarian life. Philosophers like Ralph Waldo Emerson and Thoreau worried about the human spirit itself as so-called civilization penetrated wild places, the pace of daily life accelerated, one's affairs grew in number and complexity, and people found themselves assaulted by a seemingly more mechanized world. Further, political figures in the global south such as Mahatma Gandhi deplored how industrialization was fueling colonization—turning the developing world into a vulnerable source of raw materials, enabling the exploitation of large swaths of humanity, and bringing the economistic logic of the global North to the South.[2]

Although not unified behind a single message, such thinkers were voicing a shared concern about the excesses of industrialization, and in doing so, represent the first collective expression of an environmentalist sensibility. This first expression was not, to be sure, an ideological bloc, singular school of thought, or coherent movement. Rather, it was a diverse set of criticisms leveled against an emerging array of social forces associated with the Industrial Revolution. As such, environmentalism was a set of social and moral criticisms. Romantic poets, transcendental philosophers, Victorian and gothic novelists, and other intellectuals recognized that the technological, economic, and political changes that industrialization was bringing in its wake threatened the quality of life for humanity and the natural world itself.

Environmentalism's early years set down a touchstone of what was later to become the modern American environmental movement. This is the idea of limits. Emerson, Thoreau, Wordsworth, Gandhi, and others saw industrialization as a powerful force that threatened to deface beautiful landscapes, obliterate traditional ways of life, and reformat the way that people think about and experience the world. Their environmentalism, as it were, was a matter of trying to hold back, resist, or otherwise minimize the life-changing power of industrialization. They sought to preserve wild places, rural lands, the values of more simplistic ways of life, and the kinds of experiences that nature offers for transcending the self and its urban-inspired preoccupations. Put differently, they tried to draw a boundary around the engines of industrialization while protecting the rest of the lifeworld and natural environment from being so colonized.

A large part of their motivations stemmed from the fear that industrialization would finalize what I will later describe as modernity's appetite for mastery over the wild and unbidden character of the world. They railed against this. They sought to preserve the untamed, the feral, and the rawness of life. As Thoreau writes, "All good things are wild and free."[3] The Industrial Revolution threatened such things with its mandate to convert wildlands into sources of raw materials, standardize production, and create reliable networks of distribution. Early environmentalist inklings involved erecting a conceptual boundary, policing the barricades, and trying to set limits on the mastering impulses of humanity. On one side of the divide was nature or at least nature as largely unformatted by humans—the wild, self-willed, or other-than-human-willed world that operates independent of human intention. On the other side was humanity—the tamed realm of human artifact, culture, and sociability. Early environmentalists sought to circumscribe and protect the former. Limits were a form of cordoning off and consecrating the wildness of the earth.

While not a movement per se, these early voices spawned activist efforts. They gave way in Britain, for example, to the

Commons Preservation Society, the Lake District Defense Society, and arguably the most influential, the National Trust. In India, Gandhi's efforts, which to be sure were not simply environmental in spirit, encouraged a huge voluntary simplicity movement dedicated to the values of rural and village life with organizations such as the All India Village Industries Association and the All India Spinners' Association. In the United States, Emerson and Thoreau influenced naturalists such as Muir and politicians like Theodore Roosevelt who helped form groups like the Sierra Club and the Boone and Crocket Club, respectively.

Environmentalism's early years were marked not only by romantic intellectuals and activists fronting an artistic or affective response to industrialization but also by social and natural scientists who worried less about the aesthetics of rural life and the moral fiber of society, and more about the ability of nature to provide enough resources, sinks, and ecosystem services to keep society going. More empirically minded, this second group of thinkers erected a different set of boundaries. Their voices, and the borders they fought to establish and protect, have nonetheless also become central to contemporary environmentalism in general and American environmentalism in particular.

To this second set of thinkers, industrialization threatened not simply the quality but the viability of life. Nature was not just a place that offered unique experiences or harbored the essence of wildness but also a set of biophysical entities and relationships on which humans fundamentally depend. Nature provides resources, in the form of food, wood, fuel, and minerals, and absorbs biological and technical waste. To many, industrialization seemed to have an endless appetite for natural resources, and with the harnessing and burning of fossil fuels and industrial-size production processes, was generating a seemingly incessant amount of waste. It was gobbling up forests and mineral-rich lands, pumping exceeding amounts of industrial by-product into the earth's waterways, air, and soil, and in the process, compromising the ecosystem services of the earth. Some people started to realize that such a system was unsustainable. The earth is capable of producing only so many resources and only at a certain pace, and can absorb only so much environmental assault. An insatiable appetite for resources and an out-of-control production process would eventually run up against limits. To this second set of thinkers, these limits were not ideational but physical. They involved matters not of ideals but survival.

One of the first and most forceful voices to raise the specter of sustainability and set into motion broad concern about the biophysical limits of nature was the political economist Thomas Malthus. In his often-referenced *Essay on the Principle of Population*, which Malthus continually revised between 1798 and 1826, Malthus worried that the gifts of industrialization, including advances in medicine and sanitation, were enabling people to live longer, have more children, and reduce infant mortality. He predicted that this would create significant problems as accelerating growth in population would eventually outpace increases in food production. Malthus foresaw a world of much misery as people would begin to fight over dwindling food supplies, pestilence would spread as a result of unhealthy living conditions, and famine would visit increasingly larger parts of the world. To Malthus, the earth had only so much fertility to it, and while this could be unlocked and partially taken advantage of through human ingenuity as well as enterprise, it could only be done at a certain pace and up to a definite limit. At some point—Malthus predicted the middle of the nineteenth century—humanity would rub up against the limits of material productivity and thus face dire consequences.[4]

While Malthus worried about food, at a higher level of abstraction he was concerned with material limits in general. He envisioned a line between natural and social systems, and feared what today we would call ecological "overshoot" as the latter overwhelms the former. His thought has served as a foundational message for much of the environmental movement, especially in the United States. After Malthus, people started to see multiple edges or boundaries that—no matter how brazen humans were—could not be crossed without harmful consequences.

Nineteenth-century scientists, for instance, began worrying about wood, water, and wildlife in the world's forests as the Industrial Revolution whipped across the planet, and deepened its reach within Europe and the United States. Throughout the colonies, rain forests were converted to tea, cotton, and sugarcane plantations to feed the growing appetites of those in the metropole. Within the United States and Europe, the effects of industrialization on forests were equally felt as railroads opened up whole new tracts of land and intensified ongoing deforestation. Concerns about these took on a Malthusian tone as nineteenth-century scientists like Alexander von Humboldt and Dietrich Brandis of Germany witnessed intense deforestation in South America and South Asia, and warned that while cropland is important, such deforestation robs people of needed firewood, erodes the land, and diminishes fresh water as trees no longer provide shade, hold moisture, or otherwise protect creeks, rivers, and lakes from wearing away. Similar thinkers in various parts of the world expressed related fears as they saw vast tracts of verdant land fall under poor cultivation methods and suffer the consequences of unmindful resource use. These people understood that you can push nature only so far. At some point in time it will buckle under too much assault, and when this happens, the foundations that keep life alive will falter.

Boundaries and the Birth of Modern American Environmentalism

The American environmental movement has always been informed by both the romantic and scientific wings of eighteenth-

and nineteenth-century thought. On the romantic side, later environmentalists shared the view of Wordsworth, Blake, Thoreau, and others that wildness is something to be prized and protected for aesthetic, spiritual, and moral reasons. Naturalists like Muir, Roosevelt, and Leopold, for instance, saw wildness threatened by industrialization's rationalistic, mechanistic, and homogenizing mentality, and sought to protect it by building and policing a boundary between civilization and the wild. They helped form a number of wilderness groups in the early to mid-twentieth century—such as the National Audubon Society, Wilderness Society, and National Wildlife Federation—and used these to translate the boundary into national parks, wilderness areas, and wildlife preserves that draw actual lines in the ground aimed at protecting nature's wildness from the encroachment of civilization. This tradition of thought and concern continues to course through the American environmental movement. It constitutes what many see as the preservationist wing of American environmentalism.

Likewise, a tradition has grown out of the work of Malthus, Humboldt, and Marsh focused on sustainability. Gifford Pinchot, the first secretary of the U.S. Forest Service, Henry Adams, historian and grandson and great-grandson of presidents, and Harold Ickes, the interior secretary under Franklin Roosevelt, for instance, all embodied this trajectory in their focus on the scale and pace of resource extraction and waste production. This wing of the movement developed the so-called gospel of efficiency, which counsels wise and careful use of nature's bounty rather than the unbridled consumption of it. Thinkers and activists of this ilk refuse to draw a hard-and-fast line dividing nature and humans, or wildness and civilization, and seek instead a dynamic boundary across which humans can travel, but importantly, *only up to a point*. At some point, too much meddling with nature will undermine the land's, water's, or air's ability to provide resources or absorb waste, and humans will suffer as a result. Such advocates of sustainability set the stage for groups like the Clearwater Project, Zero Population Growth (which is now known as the Population Connection), and the Natural Resources Defense Council, which caution against the unrestrained, rampant, and exploitative treatment of the natural world. The modern American environmental movement emerged as an amalgamation of these two streams of thought.

While American environmentalism can be traced back to the eighteenth and nineteenth centuries, and while it gained greater sophistication and development throughout the early twentieth century, it really came into its own in the 1960s and early 1970s. Many locate its birth with the publication of biologist and nature writer Rachel Carson's *Silent Spring*, and the general demeanor that the book helped launch. *Silent Spring* describes the ecological effects of widespread and growing pesticide and herbicide use. It shows how the chemical revolution of the 1950s, which was simply an extension of the Industrial Revolution more generally, was contaminating the earth's air, water, soil, and species, and laments the illnesses and biological diminishment that the chemical revolution brought in its wake. When it was first published, *Silent Spring* sounded the environmental alarm as it explained how industrial pollutants were literally poisoning the environment as well as endangering humans and all forms of life.

Carson dramatized environmental dangers by writing about them in warlike terms. To her, the sheer amount and extensiveness of poisons poured on to and throughout the earth resembled a battle between humans and the natural world. She worried that humans were winning, and in turn, "changing the very nature of the world—the very nature of its [the earth's] life."[5] Carson drew connections between the synthetic chemical industries and World War II weapons production, and saw pesticides and herbicides as part of an arsenal of "elixirs of death" that included, among other dangers, radiation from nuclear testing. Carson observed, "The question is whether any civilization can wage relentless war on life without destroying itself, and without losing the right to be called civilized."[6]

Carson wrote about the war against life as taking place across the old human/nature boundary envisioned by Marsh, Humboldt, Brandis, and Pinchot. It was a matter of overextending ourselves into the nonhuman world. Indeed, Carson did not reject pesticides or herbicides outright. She believed that they could be used in small, targeted quantities to destroy weeds, protect crops, and assist in diminishing insect-borne diseases. Rather, she objected to farmers, gardeners, and municipal arborists indiscriminately blanketing fields, parks, and college campuses, thereby overwhelming the earth's ability to dissipate, break down, or otherwise neutralize such poisons. She understood that the residues of herbicides and pesticides would bioaccumulate up the food chain—killing worms, insects, and birds—and disrupt the healthy functioning of creatures (including people) and ecosystems in general. Carson identified a boundary beyond which she thought human beings should not go. We can interact with the natural world: use its resources, count on it to absorb our waste, and enjoy the recreational and aesthetic opportunities that it offers. Yet if we delve into and disrupt its workings too deeply, as she envisioned we were in our chemical war against nature, we alter its reproductive, chemical, and biological abilities. For Carson, doing so is not only unwise but ultimately suicidal.

Silent Spring remains one of the most forceful warnings about human intervention into natural processes. At the time it was published, it started a virtual revolution in thought, and was soon followed by other clarion calls of concern. For example, people like Stanford University biologist Paul Ehrlich presaged global ecocide as human population would grow beyond the earth's carrying capacity. In *The Population Bomb*, he warned that the "current rates of population growth guarantee an environmental crisis which will persist until the final collapse."[7] The population bomb, according to Ehrlich, would go off as human numbers exceed the biophysical limits of the earth. Like Carson, Ehrlich recognized real material limits across which humanity cannot tread with impunity. Similar warnings came out of the Club of Rome report—written by a group of systems analysts, natural scientists, and engineers, and published as *The Limits to Growth*—which envisioned burgeoning population and consumption pressing the generative capacity of the earth to produce resources and absorb waste. Like Ehrlich, the book's authors recognized thresholds across which humans dare not go. They foresaw a pattern of growth in almost all dimensions of the human world, and saw this in conflict with the finitude of the natural one. They thus predicted that the clash would lead ultimately to a "collapse into a dismal, depleted existence."[8] An updated version of the report confirms this prognosis.[9]

Carson, Ehrlich, and the authors of *Limits to Growth* represent the tip of an iceberg of cultural concern for ecological well-being. Throughout the 1960s and 1970s numerous books, articles, films, and popular songs alerted society to the dangers of ecological overshoot and planetary fragility. Moreover, American society began actually witnessing and experiencing resource scarcity and environmental harm throughout this same period. In the early 1970s, for example, in response to U.S. support for Israel in the Yom Kippur War, the Organization of Petroleum Exporting Countries (OPEC) restricted production and raised the prices of petroleum, leading to lines at the gas pump and price increases throughout society as the costs of using oil went up. This "energy shock" made many Americans aware of resource scarcity seemingly for the first time. Likewise, a number of events brought the dangers of pollution into high relief. A huge oil spill off the coast of Santa Barbara, the bursting into flames of the Cuyahoga River, the nuclear meltdown at Three Mile Island, and reports of contamination of Love Canal underlined the stakes of ill-managing industrial society and the problems associated with exceeding amounts of waste. These events, and the gathering sense that resource scarcity and pollution were compromising the quality of life, inspired the emergence of groups such as Greenpeace, the Union of Concerned Scientists, and Friends of the Earth, many with central offices in the United States. In short, a "cataclysmic temper" descended over America throughout the 1960s and 1970s, giving powerful expression to environmental concerns.[10]

This temper extended beyond resource and waste concerns to wildlands as well. Following in the footsteps of Wordsworth, Thoreau, and Emerson, and later Muir, Roosevelt, and Leopold, quite a number of thinkers and activists arose in the late 1960s and 1970s to protect natural places and nonhuman creatures. This wing of the movement continued to insist that wild places have not only biological significance but also moral, spiritual, and aesthetic value. This dimension of the movement gave voice to animal rights, the importance of biological diversity, and the notion of intrinsic value to things other than humans, and helped establish organizations such as the Nature Conservancy, World Wildlife Fund, Conservation International, and Earth Island Institute. Like other dimensions of the modern American environmental movement, it identified the divide between humans and nature, and tried to staff the barricades between them.

Conclusion

The environmental movement has changed in many ways since its early days in the eighteenth and nineteenth centuries, and its modern incarnation in the 1960s and 1970s. It has grown tremendously in size and geographic reach, and enlarged its agenda to focus on everything from wilderness protection, toxics, and climate change, to social justice, peace, and indigenous people. Moreover, the movement has become mainstream; where once environmentalism was a lone voice crying out to a deaf public, it is now a central expression of American society. While the movement's size and reach along with the palatability of its message have shifted over the last few decades, remarkably its approach to understanding and responding to environmental issues remains relatively unchanged. The American environmental movement is still fundamentally animated by a set of principles, emerging out of its early years, which guide its sensibilities, frame its policies, and in general orient it to the broad task of pursuing environmental well-being. At the heart of these principles is the human-nature divide—a distinction that implicitly calls on people to preserve, conserve, and sustain the viability, functionality, and wildness of the earth.

Environmentalism has offered much to the world by analytically separating humans from nature and recognizing how much the latter means to us. American environmentalists have won many victories and advanced the cause of environmental protection based on the dichotomy. In this chapter, I have briefly traced the evolution of the distinction as it emerged within the American environmentalist tradition. This hopefully helps one appreciate the long historical reach that the concept of nature has had in the movement and the political efficacy of distinguishing nature from humanity.

Boundaries do not simply separate but also encourage people to privilege one side of a divide over another. By separating nature from humanity, early environmentalists did not merely draw analytic distinctions but left their descendants a legacy of preference as well. They implicitly called for later environmentalists to choose the realm of nature over that of humanity and work on behalf of nature in contrast to the human-made world. In the following chapter I explain this legacy, and what it has come to mean philosophically and politically for the movement. I do so by explicating the dream of naturalism and the role nature has played and continues to play in the American environmental movement.

"Circle," Cape May, New Jersey

3

The Dream of Naturalism

Our views of nature imply answers to questions about the very meaning of life.

—Kerry Whiteside, *Divided Natures*

When modern American environmentalism was emerging in the 1960s, many young people from the cities started moving out to the countryside. Disillusioned by the glitz of urban life, they built farms, grew their own food, and attempted to reconvene with nature. The back-to-the-land movement, as it became known, sought a way of cutting through the affectations of the times, digging below the surface of social cues, and discovering a deeper sense of self and life. In nature, the back-to-the-landers found what they thought was a route to more authentic living. Nature, in all its seeming purity and nakedness, appeared as an uncontaminated realm in which one could experience life unadorned by the corporate, consumer-based pretentiousness that characterized American society at the time.[1] While the back-to-the-landers never made up more than a fraction of the environmental movement, they expressed a type of love of, trust in, and respect for nature that has always been implicit in the movement's long history, and continues to animate environmentalists to this day. They recognized that the green "grit" of the earth represents not only a route to ecologically safer living but also a "good" that should be pursued and heeded if one is to live a worthwhile life. They saw the good life epitomized by living close to nature, and in conformity with its ecological imperatives and limits.

The back-to-the-landers did not, of course, invent the valorization of nature or the search for authenticity. The idea of living according to nature to protect one's health and safety, and as a means to a good, fulfilling life, has resonated within various historical movements that reject contemporary culture in search of a more secure and genuine existence by looking to nature for inspiration. Nature, in this sense, has long been associated with all that is good, beautiful, true, and right in the world. As John Stuart Mill writes, "That any mode of thinking, feeling, or acting is 'according to nature' is usually accepted as a strong argument for its goodness . . . and the word unnatural has not ceased to be one of the most vituperative epithets in the language."[2] The back-to-the-land movement translated the urge to follow nature into concrete form. By returning to the land, and dedicating themselves in both thought and action to nature, they showed that there is something fundamentally attractive about the natural world.

What is the urge toward nature, and how does it inform American environmentalism? How does the love of, trust in, and respect for nature animate the thought and politics of the movement? In the following, I explain that nature serves as an object of aspiration for many environmentalists. It is a good toward which a sizable number of American environmentalists gravitate. The back-to-the-landers did not melt into the landscape when they left the city; they did not *become* nature. Rather, they oriented their lives toward an ideal of harmonizing with, rather than imposing themselves on, nature. Nature, as such, disciplines practice and provides a direction for policy. It serves as a guide for negotiating one's way in the world. Many environmentalists use it as a trajectory for thought and action. In doing so, they embrace what I have been calling the dream of naturalism. Naturalism is the notion that we live best when we align ourselves with the natural world, when we take our cues from and otherwise orient ourselves with respect to nature. It is a dream insofar as we never fully integrate into the natural world but nonetheless maintain the desire to do so.

The dream of naturalism is not simply the sensibility at the heart of environmentalism but also represents one of the poles in environmental politics. As mentioned in the introduction, environmentalists and their critics square off over the place of nature in our lives. While many environmentalists subscribe to the dream of naturalism, environmental skeptics aspire toward the dream of mastery. Skeptics see human beings as superior to the natural world, and thus believe that humanity can and should outsmart, manipulate, and ultimately subdue nature in the service of human betterment. The dual dreams of naturalism and mastery have polarized environmental politics for decades, and continue to do so. Coming to terms with the dream of naturalism, then, provides not simply a deeper understanding of American environmentalism but also helps articulate the rifts within environmental politics and opens lines of thought for envisioning a postnature environmentalism.

It is essential to note one thing before proceeding. At the beginning of the last chapter, I mentioned the variegated character of the American environmental movement. I explained how light greens disagree with dark greens, grassroots groups take issue with national organizations, and many wings of the movement are animated by issues that only tangentially have to do with protecting wildness. It was only after noting this that I identified the dominant line of thought and practice in the movement that divides humans from nature, and tries to protect the wildness of the latter from the former. The same qualification is relevant for this chapter. In the following, I will suggest that deep down, most American environmentalists subscribe to the dream of naturalism. Most recognize the necessity and desire to harmonize human life with the natural world. In asserting this, I do not want to imply that this is true of all

American environmentalists or the movement qua movement. There are many thinkers, activists, and organizations within the movement that approach environmental challenges from a different orientation. Nonetheless, the dream is real enough and pervasive throughout the movement. The American environmentalist imagination has long seen nature as representing much that is true, good, right, and beautiful in the world. Appreciating this is important, since prizing nature as such makes nature's dismemberment and the taming of its wildness by both our minds and bulldozers so scary.

The True: Prudence and Biophysical Limits

"Nature bats last" is a familiar environmentalist slogan. It means that no matter what human beings do, sooner or later the laws of nature will express themselves and take precedence over human activity. For example, people can build houses in floodplains, pump excessive amounts of carbon dioxide into the air, or wipe out species at inordinate numbers. At some point in time, however, human activity will be vulnerable to the patterns of nature—patterns that are blind to the intentions or well-being of human beings. Environmentalists tout the slogan to remind people that while humans can try to control nature– and have done so in truly remarkable ways–ultimately nature is sovereign. The laws of physics and chemistry, at the heart of nature's dynamics, will take precedence over whatever designs humanity may dream up. Ignoring nature's sovereignty, then, is folly: it imperils human life, and the relative stability and health of the earth's ecosystems.

Behind the idea of nature's sovereignty is the notion of necessity. Nature "rules" because it represents that which cannot be otherwise. It is the "given" rather than the "made," and operates according to preordained patterns that are indifferent to human life. As a matter of necessity, nature can be manipulated—and indeed always is, as humans interact with it—but its ultimate physicality and fundamental structural characteristics are not up for negotiation. One either respects the laws of nature or is blindly at the mercy of nature's power. In the end, nature sets the limits of human activity. When nature's thresholds are crossed, populations crash, landscapes are defaced, air is polluted, and climate changes. People can, of course, come to appreciate or at least tolerate these changes, but the ultimate authority is nature itself. In such cases, nature has "spoken." It is nature whose parameters have been pressed and whose dynamics have expressed themselves.

Today nature is speaking rather loudly. For instance, the buildup of greenhouse gases in the atmosphere represents the planet saying that it has reached its limit for absorbing carbon dioxide, methane, and other greenhouse gases. After we have pumped such gases into the air for a couple of centuries—expecting the oceans, forests, and atmosphere to soak up or otherwise neutralize them—the planet seems to be saying, "I have had my fill. I can absorb no more." What is going on, according to environmentalists, is that humanity has encroached too deeply into the natural world, and the latter is drawing a line, proclaiming, "Here, no further." The same dynamic is going on with ozone depletion, biological diversity, and freshwater. Humans have released an overabundance of chlorofluorocarbons and other ozone-depleting substances into the atmosphere, denuded too many plant and animal habitats, and polluted too many streams and rivers, thereby crippling the earth's ability to retain the stratospheric ozone layer, maintain biological diversity, and filter and regenerate freshwater. In these cases, the sovereignty of nature is apparent. Humans cannot continue business as usual, expecting the earth to be forever pliant.

That nature has limits is not to say that it is always in a homeostatic balance or somehow does not change. Nature, like all else, changes all the time. Creatures grow, species evolve, and ecosystems alter in the face of exogenous and endogenous forces. Nature's sovereignty simply means that nature unfolds in given ways, and that such unfolding can proceed even with human influence. At some point, however, humans will exert enough pressure to sidetrack nature's progression, and often in ways that jeopardize the quality and viability of both human and nonhuman life. This is what environmentalists mean by limits.

Preservation, conservation, and sustainability, long enshrined in the annals of American environmentalism, are principles that seek to respect nature's limits. They focus mainly on two dimensions of environmental harm. First, they are concerned with resources. The earth produces, by its very aliveness, a vast array of animals, plants, and minerals that we use to feed, shelter, and fuel ourselves, and that the earth cycles through its systems to maintain ecological health. All of these are renewable in the sense that they regenerate over time; the earth can replenish resources. The problem is that the natural rate of regeneration for most resources is slower than the speed at which humans are consuming them. Timber, fish, topsoil, and freshwater renew themselves relatively quickly; oil, coal, and various minerals obviously regenerate at much slower rates. Huge increases in population, affluence, and technology have allowed us to use both sets of resources at an accelerating pace—putting tremendous pressure on the earth's ability to sustain its "natural capital." This makes it difficult for the earth to continually produce raw materials for human use and the well-being of other creatures.

Depletion of resources has long preoccupied various wings of the environmental movement. As mentioned, political economist Malthus started the discussion by worrying about the earth's ability to produce increasing amounts of food to feed a growing population, and many others have continued the discussion, citing the depletion of various resources—fertile soil, fish, minerals, and freshwater—central to contemporary societies. Today, concern about resources still animates the environmental movement. We know, for example, that two-thirds of all fish stocks in the oceans are overfished, and catches have been declining for much of the last decade.[3] Likewise, overharvesting from aquifers has led to a freshwater scarcity crisis in which one-third of humanity lacks access to safe drinking water and another third enjoys only intermittent freshwater supplies.[4] Additionally, petroleum use over the last few decades has brought us to a time of "peak oil," where we have passed the point of maximum production such that growing demand will soon easily exceed supply.[5] Each of these concerns recognizes that demand is outpacing supply in many natural resources.

Environmentalism has always concerned itself not simply with resources but also with sinks. Every process of production and consumption generates some type of waste. The earth has the capacity to absorb much of this waste without compromising its ability to provide ecosystem services or simply maintain ecosystem health. The problem is that humans often overwhelm the earth's ability to do this. The earth can neutralize waste only at a certain rate. When we dump too much by-product into our water, air, or soil, the sink stops up. This is why environmentalists worry not simply about running out of raw materials but also about the waste stream that accompanies the extraction, processing, distribution, and consumption of materials we already have at hand. So while environmentalists fear peak oil, they also worry about the risks associated with waste generated by actually burning known oil reserves. These risks include increased air pollution, smog, and the dangerous buildup of greenhouse gases. Similarly, environmentalists worry about not only deforestation but also the waste stream (dioxins and other effluents) generated by the process of turning timber into processed materials like paper.

Overloading sinks, like overusing resources, has been a long-standing concern. Charles Dickens wrote of fouled air, Carson about the bioaccumulation of harmful pesticides, and McKibben of the buildup of greenhouse gases. These kinds of thinkers and activists see nature not merely as a provider of raw materials but also as a gigantic cleanser that purifies and re-generates air, water, and soil. When we undermine nature's cleansing services we begin to choke or otherwise contaminate ourselves and other beings. Put differently, nature is not in-finitely malleable or exploitable when it comes to the earth's cleansing abilities. Our overgeneration of waste fouls not sim-ply our nest but the nests of every creature on earth as well.

That nature operates according to necessity and abides by inalterable laws is the reason many environmentalists see it as the "true." We can entertain various ideas about how the world is and should be. We can also act in various ways indifferent to nature's laws. At some point, however, nature's sovereignty will make itself known; at some point, the invariant, concrete actu-ality of nature will reveal itself. And when it does, we witness something veritable about the world. Nature is not something imaginary but rather something genuinely real. Indeed, because it sets the biophysical constraints on life, it represents the most real, as it were. It thus becomes something on which environ-mentalists ground their perceptions of the world. Nature is the most certain of things.

The Good: Biomimicry

The sovereignty of nature is a powerful idea. It means that something else, something other than humans, sets the condi-tions for human life. No matter what we think or want, the natural world will unfold not as we would necessarily like but rather as it needs to. That nature operates according to neces-sity leads environmentalists to endow it not only with empiri-cal weight but also normative significance. The natural world not only operates in a certain fashion but seemingly *should* operate in this manner, and we would do best to follow its cues. Here the idea is not that we must fear nature because it is an unforgiving force but that we should actually emulate it as a standard for living authentic, spiritually rich, and ethi-cally upright lives. Known philosophically as *naturam sequi*, this orientation sees nature as offering maxims for living well. We thus follow nature not because it threatens our biophysical well-being per se but also because doing so offers a more meaningful life.

At the heart of *naturam sequi* is the view that nature is of a higher dignity than human life. The ancient Greeks distinguished between *physis* (nature) and *nomos* (law, custom, or convention). They did so to differentiate the "given" from the "made." The former is unconnected to any consciously purposeful activity—and hence fundamentally independent of human agency—while the latter is a product of human thought and purpose.[6] Many thinkers throughout the ages and many contemporary environmentalists privilege physis because it seemingly expresses the way of the universe. It represents the way things would be if human beings were out of the picture. It therefore has a primacy to it that is missing from human laws, customs, and the like, because the latter are matters of human interpretation and sociality rather than characteristics inscribed in the very essence of the world. What is curious is that physis enjoys this primacy not simply because nature seems to operate according to necessity—where one must respect nature *or else*—but rather because physis is unblemished by the imperfections and incongruities that often mark human life, and thus stands as a model for those seeking to live in the highest ways possible. Put differently, *naturam sequi* represents the age-old aspiration to seek extrahuman sources of value, inspiration, and philosophical foundation for human life. One finds it, for instance, in Cicero when he announces, "I follow the guidance of Nature. . . . Not to stray far from Nature and to mould [*sic*] ourselves according to her law and pattern—this is true wisdom."[7] Environmentalists have long subscribed to this view. Many of them believe, as biologist Barry Commoner puts it, that "nature knows best."[8] This means that it is not only prudent but also pedagogically significant to follow nature's dicta.

Environmentalists have taken this in multiple directions. In its most ideological form, it leads to something that

British political scientist Andrew Dobson calls "ecologism."[9] This is a commitment to see the natural world as a model for the human one, and arrange our political, social, and economic lives accordingly. For example, various schools of environmental thought such as social ecology, bioregionalism, and ecoefficiency advance a type of "ecocracy" that calls on humanity to build the characteristics of nature into our social practices. Social ecologists, for instance, argue that nature works according to principles of equality (not hierarchy), differentiation (not uniformity), and cooperation (not competition). They thus call on humans to fold these same principles into our institutions and personal relations.[10] Bioregionalists insist that humanity is currently organized around political boundaries rather than ecological ones, making it difficult to live ecologically sensitive lives. They advocate that we study nature's systems to gain direction for organizing human settlements, and instill a sense of ecoawareness and appreciation for place in our lives.[11] Those who espouse the principles of efficiency and recycling do the same thing as they claim that these principles are part of nature's ways, and that we should imitate them in our collective lives.[12]

Behind all these ideas is the notion that the human order should be based on the natural one, that we should turn to nature for cues to good living. Political theorist John Meyer sees such an orientation running through much Western political theory in general and calls it a "derivative" view of politics insofar as it develops normative principles based on naturalistic understandings.[13] In other words, we would best flourish if nature were the source of our political, social, and economic lives. To many environmentalists, this is possible if and when, as farmer, novelist, and philosopher Wendell Berry puts it, "we can find the humility and wisdom to accept nature as our teacher."[14]

The idea of *naturam sequi* is so prevalent among environmentalists that almost all American environmental advocates advance some version of it. Early environmentalists like Muir,

Bob Marshall, and Leopold certainly believed that the nonhuman world had much to teach humans, and that we should look to nature for healthy, right living. Contemporary environmentalists such as Carl Pope of the Sierra Club, David Suzuki of the David Suzuki Foundation, and Dave Foreman of the Wildlands Project also counsel turning to nature for cues about how best to live. Curiously enough, it is not simply the nature lovers who express a type of *naturam sequi*. Today, the most technologically minded environmentalists are also seeking fundamental insight from nature. Architects such as William McDonough are designing buildings that work like trees—turning solar energy into power, neutralizing waste, and providing habitat for diverse species.[15] Agriculturalists like Wes Jackson are developing ways to farm "in nature's image." They are breeding perennial crops that can grow in untilled soil and thrive with little human input.[16] Even business leaders such as Paul Hawken and Ben Cohen are using nature's processes to organize corporate practices, emphasizing principles such as mutuality, symbiosis, efficiency, and waste reduction.[17] Author Janine Benyus labels these efforts instances of "biomimicry," and explains their potential for reorienting the way we grow food, produce energy, make products, heal ourselves, and conduct business. She tracks such attempts to show that following nature will not only protect us from grave environmental dangers but also enhance our experience of the world.[18] Humans will flourish physically, economically, socially, and spiritually in societies modeled on nature.

This normative attraction to nature differs from the prudential one in that it sees nature not strictly as a realm of necessity but rather one of choice and tendency. As the Cicero quote suggests, we have the choice to follow nature or not. This sense of choice finds its deepest roots in Aristotelian philosophy, which suggests that all things have a nature about them—that there is an essence to things that defines and constitutes them—and that a thing flourishes best when it realizes its true nature. Aristotle offers multiple explanations of a thing's nature, referring at times to an entity's form, origin, matter, source of movement, or end toward which it strives (*telos*).[19] What is most relevant is that humans also have a nature or essence, and realizing it takes insight and effort. For environmentalists, we can fully realize our essence by modeling our lives after nature. In its grandest sense, this involves formatting society and human practices on ecological principles—an enterprise that privileges the natural over the artificial, the given over the made. To be sure, not all environmentalists share this sensibility, but the idea that human life is best lived by following nature is central to many manifestations of American environmentalism.

The Right: The Earth Community

Looking to nature as a guide for living is a matter of principle. It involves privileging nature and choosing to organize one's life accordingly. Yet many American environmentalists embrace nature as a matter of not only principle but also moral practice. This involves not emulating nature per se but instead letting nature be—that is, letting it evolve or otherwise develop independent of human needs and desires. (This is one expression of the preservationist tradition in the movement.) The choice or some would say obligation to let nature be is related to *naturam sequi*, but it has a different complexion insofar as it is not about mimicking nature but rather respecting the sheer "otherness" of nature, and letting that otherness express itself. Shielding nature from human manipulation is, in other words, a moral act.

Interacting with nature is inevitable for human beings. Almost everything we do involves using nature in some capacity. Simply meeting our physiological needs requires us to use nature—and often in extensive ways. This inescapable fact of human life can have its excesses, however, and pointing this out has long been central to environmentalism's message. Environmentalism is largely about complaining that our exploitation of nature need not be as intense or expansive as it tends to be, and offering ways to restrict our excessive interventions. Advancing this message, however, has been challenging, as the viewpoint takes issue with the longstanding Western tradition of seeing nature as inferior to humans. When we see nature as less important than ourselves we take license to treat it as we see fit. For many environmentalists, this predicament is at the root of our environmental woes and thus combating it is central to environmentalist efforts.

Environmentalists give different explanations for the belittling of nature. Some thinkers identify the Judeo-Christian tradition and its insistence that God gave humans dominion over the earth's plants and animals as the source of environmental harm.[20] Others blame modernity, and its emphasis on technological vanity and the urge toward human mastery over nature.[21] Still others point to patriarchy, and its gendering and exploitation of nature, or capitalism and its commodification of the nonhuman entities.[22] In each of these accounts, the problem stems from humans seeing nature in instrumental terms. The Judeo-Christian tradition, modernity, patriarchy, capitalism, and the like understand nature as existing for humans or at least being so subordinate to us that we are given free moral reign to treat it as we deem appropriate. In the language of environmental ethics, our institutions, structures, and ideologies make us anthropocentric: they animate us to privilege human life above all else, and relate all that happens in the world to human interests and concerns. Such privileging denies any sense of intrinsic worth to plants, animals, mountain ranges, or ecosystems. These entities are not regarded as ends in themselves but rather matter only to the degree that humans can use them. As a result, humans are able to exploit nature endlessly because, from an anthropocentric perspective, nature enjoys no inherent moral significance.

Anthropocentrism finds its moral supremacy in the ethical framework of the Western philosophical tradition. Most schools of ethical thought start with the premise that morality is reserved for rational beings who can make conscious choices about their lives. That is, something must have the ability to reflect, sense its own autonomy, or otherwise reason about its existence for it to deserve moral consideration. Most of us give little thought to manipulating a car, for instance, because cars lack autonomy or a sense of self. Much of the Western tradition has looked at nature in this same way. It has seen plants, rocks, waterfalls, and animals as devoid of reason, and thus ineligible for genuine moral regard.

Different environmentalists over the ages have resisted this assessment in various ways. For example, the animal rights movement, which has played and continues to play an important role in environmentalism, has consistently argued that animals have discernible levels of consciousness that enable them to feel pain, act with volition, or otherwise experience a sense of self. The minimalist position here has been expressed by those like ethicist Peter Singer, who contends that animals are sentient beings that experience pleasure and pain, and therefore our treatment of them is a matter of moral significance.[23] Others like philosopher Tom Regan point out that while animals may not be fully rational in the ways in which many of us envision, they still retain a sense of self in that their lives matter to them. This makes them, along with children and people who are cognitively impaired, deserving of moral worth.[24] While most environmentalists are not scholars of ethical treatises, many intuitively share the sensibility expressed by the aforementioned writers, and this motivates them to work for more humane treatment of farm animals, against vivisection, in opposition to hunting, and in favor of fair treatment of non-human species. Some semblance of this sensibility underpins campaigns to save whales, dolphins, elephants, gorillas, and other mammals as well as protect against the exploitation of animals more broadly.

The animal rights movement seeks to expand the circle of the moral community to include animals. Others go further by claiming that people should extend moral consideration to all forms of life including plants and other organisms—a form of biocentrism. Biocentrism holds that human beings are not the center of the universe but simply one species among others, and therefore are undeserving of special status. Biocentrists provide various justifications for including the nonhuman world of plants, animals, microorganisms, and so forth in the moral community, and are not merely tied to Singer's criterion of sentience or Regan's subject of a life. Biocentrists believe that nonhuman entities have a right to exist simply because they are alive in the world. The aliveness of nonhuman entities endows them with intrinsic worth, and affords them rights independent of how human beings might use them, what they mean to us, or any level of consciousness they may display. Those concerned with biological diversity, habitat preservation, and wilderness protection usually subscribe to some form of biocentric thinking, and thus see environmentalism as basically a nature-centered type of moral practice. Environmentalists of this ilk are often challenged when deciding what moral consideration entails, since as mentioned, being human requires us to interact and use nature in instrumental ways. While biocentrism offers no singular hard-and-fast rules, it does provide a general receptivity to the existence of other living entities on earth.[25]

If biocentrism puts life at the center of moral concern, ecocentrism expands the moral boundaries even further to include all aspects of nature including inanimate entities, whole ecosystems, and according to some the entire earth. It celebrates the sheer otherness of nature, and argues that this in itself calls on people to treat nature with moral respect. Frequently associated with the philosophy of deep ecology, ecocentrism has inspired efforts to protect wildlands, species diversity, and all expressions of ecological abundance. It provides elements of the moral foundations for environmental legislation such as the U.S. Wilderness Act and campaigns to keep development out of places like ANWR. Biocentric and ecocentric orientations also inform some of the more so-called radical environmental groups such as Earth First! and Earth Liberation Front that fight to defend what they take to be the rights of the nonhuman world.

The rejection of anthropocentrism—whether in its animal liberation, biocentric, or ecocentric variety—revolves around criticizing the way humans place themselves at the center of things and lord over the natural world. Leopold articulated this critique over fifty years ago when he envisioned extending moral worth to the nonhuman world as simply the latest phase of a centuries-long effort to expand the boundaries of the moral community. In ancient times, according to Leopold, moral consideration was reserved for property-owning males. Over time, people started seeing slaves as fully "human" and extended moral worth to them as well. Eventually, women were welcomed into the fold as were all types of people—including children and people with disabilities—who were formerly considered somehow less than human. Leopold argues for extending moral consideration beyond the boundaries of humanity to include all entities. He does this by advancing his famous "land ethic," which as he puts it, "changes the role of *Homo sapiens* from conqueror of the land community to plain member and citizen of it. It implies respect for his fellow members and also respect for community as such."[26] Leopold's land ethic has long stood as one of the most eloquent and forceful criticisms of anthropocentrism.

At the heart of all such expressions is the idea that humans are not the only (and arguably not the most important) species or sentient organism on earth, and therefore must respect the autonomy of the nonhuman world. Environmentalist Foreman captures this sentiment well when he writes that there seems to be something fundamentally wrong about cutting down "a two-thousand-year-old redwood to build picnic tables."[27]

While most American environmentalists are probably not scholars of environmental ethics, they certainly resonate with such assertions. Their actions and explanations suggest that they intuitively sense that nature has intrinsic worth, and as such is deserving of our respect and fair treatment. To be sure, many environmentalists might argue about what fair treatment implies. Yet behind such disagreements is a general sense that nature has a right to exist and flourish independent of human beings.

Nature conceived as such has an element of "right" to it in the sense that treating the nonhuman world with respect reflects morally admirable behavior. Nature is not something irrelevant to human ethical life but rather provides a pivotal focus for cultivating moral principles for living. Nature, to put it differently, offers an ethical foundation for humanity. It is a realm in which justice and virtue can be sought after and exercised.

The Beautiful: Nature's Aesthetic

The ideas that humans can live morally upright lives by letting nature be and best flourish by emulating natural processes relate to a broader reason that environmentalists care about nature. Many care simply because they enjoy the experience of visiting or immersing themselves in nature. Nature is beautiful to many people. Natural places, exotic species, dramatic landscapes, unique ecosystems, and various soundscapes provide many with a strong sense of pleasure and well-being. This aesthetic dimension involves the enjoyment, love, and what some may call the soulful enrichment many people experience in nature. It is often underanalyzed because it is difficult to articulate why one is attracted to certain things rather than others; but this aesthetic dimension nevertheless plays an important role in

American environmentalism and environmentalism's valorization of nature.

Many people point out that widespread enjoyment of the natural world is a relatively new human experience insofar as most of our ancestors perceived nature as a frightening or at least threatening realm that had to be resisted, overcome, or mastered. In premodern times, people were frequently at nature's mercy. They had to work to secure food and shelter against the elements, and lived in a more naturalistic setting in which wild animals as well as the unpredictability of weather and disease constantly threatened human well-being. Over the last few centuries, many people, especially the wealthy among us, have mastered much of the natural world and thus have buffered themselves from nature's indifferent power. This has permitted individuals a newfound capacity to experience nature as a source of aesthetic pleasure—something they choose to enjoy rather than fear or struggle against. Furthermore, the unceasing human encroachment on and destruction of wildlands, species, and various landscapes over the past century has heightened many people's concern and appreciation for what is being lost; this too has fostered people's valuation of the aesthetic dimensions of nature.[28]

It is certainly the case that human appreciation and admiration for nature have grown more widespread over time. And yet it is crucial to remember that there have always been people who prized nature's beauty and saw the aesthetic experience of nature as a justification for its protection. The ancient Hebrew prophets and psalmists expressed wonderment at nature's beauty, as did ancient Greek and Roman philosophers and dramatists, and many of them lamented the destruction of nature on purely aesthetic grounds. This sensibility has been given a more distinct voice by modern-day American environmentalists, but its initial articulations stretch far back into history. It has coursed through the environmental tradition insofar as environmentalists have always looked to nature for recreation, artistic motivation, and even religious and/or spiritual inspiration.

The Sierra Club is one of the largest and most influential environmental organizations in the United States. Established in 1892, with Muir as one of its founders, the organization still retains his vision of environmental protection. Muir was famous for his love of hiking and exploring America's forests and wildlands. He once walked from Indiana to Florida, and after he discovered the Sierra Nevada and Yosemite Valley, immersed himself in the California wilderness on and off for years. Muir's admiration for nature stemmed from the sheer joy he felt at being in it. He was known for heading into the woods with only a few tea bags and a loaf of bread, and exposing himself to nature's rawest expressions. He enjoyed climbing trees during rainstorms and swaying in the wind. He scaled rocky mountain cliffs and would walk for days through unexplored terrain. Part naturalist, Muir studied the structure of flowers, habits of animals, and undulations of the land. He felt constantly moved by what he saw and experienced. For Muir, nature was not simply a stock of natural resources but rather a place for the most satisfying recreation and a source of rejuvenation. He counseled preserving nature in large part so people could experience the joys of being in the wilderness. Wilderness to Muir was the perfect antidote to the staid living in so-called civilization. He writes, "Thousands of tired, nerve-shaken, over-civilized people are beginning to find out that going to the mountains is going home; that wilderness is a necessity; and that mountain parks and reservations are useful not only as fountains of timber and irrigating rivers but as fountains of life."[29]

The idea that nature is a place to find rejuvenation and reconnect with our deeper selves has been widespread throughout the American environmental movement. The Sierra Club's motto emphasizes this insofar as it calls on people to "explore, enjoy and protect the planet." The club started as a group of people who liked to hike together, and it still offers extensive travel opportunities to visit wildlands throughout the world. It believes that getting people into nature is good for them, and strategically it counts on people to work for nature preservation once they have experienced the joys of hiking, biking, kayaking, or otherwise immersing themselves in the natural world.

Appreciating the aesthetic and rejuvenating qualities of nature is apparently not only a good to be enjoyed but also a psychological necessity. Author Richard Louv points out, for instance, how increasing numbers of children around the world, but especially in the United States and Europe, no longer experience nature in a day-to-day sense. A combination of parental fear, the disappearance of green spaces, and the increasing allure of video games, computers, and other indoor entertainments are minimizing the amount of time kids are actually "outside." The result is what Louv calls "nature-deficiency disorder," which presents itself in high levels of distractibility and even depression. To Louv, the aesthetic dimension of nature is not some luxury that we can choose to experience but is instead essential to our psychological health.[30]

Many environmentalists find that their experiences of nature are not simply aesthetically pleasurable but also philosophically and spiritually profound. Like Muir, many see nature as providing a place that strips one of society's accoutrements, thereby allowing one to feel the marrow of life that much more immediately. One of the most eloquent to express this sentiment was, of course, Thoreau. Thoreau was not an environmentalist in the modern sense of the word. As mentioned, he lived before there were organized groups concerned with environmental protection and before a set of coherent ideas could be articulated on behalf of nature protection. Thoreau's contribution to environmentalism, though, was significant as he linked experiencing nature with questions of how best to live a meaningful life. He writes in *Walden*, "I went to the woods because I wished to live deliberately, to front only the essential facts of life, and see if I could not learn what it had to teach, and not, when I came to die, discover I had not lived."[31]

This view shares much with the idea of *naturam sequi*. It differs, however, because, as Thoreau himself makes clear, it looks to nature for a certain type of experience rather than a model of how to live all the time. Although Thoreau lived at Walden for two years, he eventually left to live among his friends and family in Concord, Massachusetts, and even left Walden occasionally while in residence. For Thoreau, nature fundamentally represented a realm of spiritual evocation. Like his colleague Emerson, he believed that the divine spirit is present throughout nature and that humans can more fully experience the divine within themselves when they have intimate contact with the natural world. Thoreau feared that an increasingly mechanistic, materialistic, and urbanized environment was enfeebling humanity's spiritual essence by closing off access to the untamed, unworked, "given" quality of the world. The wild represented the route back—to one's higher self and toward what Emerson called the "Over-Soul," which permeates the entire universe. For Thoreau, Emerson, and other transcendental-minded philosophers the pull toward divinity was primarily aesthetic. In Emerson's essay "Nature," he constantly refers to the inexplicable human attraction to beauty, and how it is this impulse that propels people toward nature and thus allows them to more easily experience divinity.[32]

The idea that nature represents an essential route to the divine has a long history. It is probably no accident that the stories of many religious figures relate how such people experience their most profound insights in the wilderness. For example, we are told that the patriarchs of the three Abrahamic faiths—Moses, Jesus, and Mohammed—each felt the presence of God in the wilderness. Likewise, the Buddha gained enlightenment under the bodi tree, and Confucianism, Jainism, and Taoism have always encouraged practitioners to experience nature as a form of religious devotion. There are, of course, huge complexities when it comes to religiosity and nature—least of which is the privileging of humans over nature in many faiths, and the consequences this has for environmental stewardship. The important point is that the relationship between nature and divinity, especially because of the sublime qualities many find in nature, has long informed environmentalism and continues to do so. Muir found the thoughts of God etched into the lines of flowers;[33] today environmentalists talk of the Everglades as the "Sistine Chapel of wetlands"[34] and the Arctic National Wildlife Refuge as the "sacred place where life begins."[35] What lends this spiritual dimension to nature is that people find that encountering the beauty of the natural world moves them at a deep level.

A final dimension of this aesthetic sense is offered by sociobiologist Edward O. Wilson in his concept of biophilia. Wilson claims that each person is hereditarily predisposed to feel a deep affinity with nature. This comes from spending the bulk of our evolutionary past on the African savannah and in contact with other species. According to Wilson, at some level everyone derives a sense of profundity and aesthetic pleasure by experiencing certain landscapes and interacting with other species. This translates into an affinity for nature that may not be evident in everyone all of the time but is nevertheless ingrained into humanity's genotype.[36] There is much controversy over Wilson's idea, and he himself refers to it as a hypothesis. For our purposes it is less a matter of whether the insight is true than understanding what it implies. The biophilia hypothesis seeks to explain why so many of us feel deeply drawn to nature. For Wilson, it is a matter of genetics; for others, it is a question of historically and socially acquired tastes. Either way the intent is to explain the appeal nature has for humans and further illustrate the important role of nature in environmentalism.

In 1973, physicist Martin Krieger published an article in *Science* titled "What's Wrong with Plastic Trees?"[37] In it, he suggests that our taste for natural things may be variable, and that with large-scale environmental degradation taking place, we should perhaps develop an aesthetic sense for the synthetic world rather than the biological one. Noting that so much of the so-called natural world is already human designed—he was thinking of the parks of landscape architect Frederick Law Olmsted—Krieger recommends that we could come to appreciate, find genuine meaning in, and experience authentic pleasure in landscapes that are artificially created rather than encountered. Environmentalists reacted strongly to the article and its suggestion. Although they could not articulate it in an abundantly clear way, many felt that there was something fundamentally wrong with creating forests of artificial trees even if, from a visual perspective, such forests might be indistinguishable from the original ones. This sense of wrong is primarily aesthetic. An artificial world just doesn't sit right with many environmentalists. Nature, unto itself, carries too much weight for environmentalists. Deliberately getting rid of it thus appears as a sacrilegious act.

Conclusion

The idea that humans should follow nature is ancient. It has to do with the general notion that we should look outside ourselves, toward the more-than-human world, to grasp the limits of human possibility as well as find the source of wisdom, ethical instruction, and beauty. In this chapter, I have been suggesting that this sensibility can be understood as naturalism. Naturalism recommends that we align ourselves with, rather than impose ourselves on, the natural world. The natural world is seemingly superior to or at least sufficiently separate from the human one, and this is why it offers insight into human life. It is uncorrupted by the contingencies and interpretative plurality of human life, and thus represents a trustworthy ground for informing our lives and societies.

Understanding nature's imperatives, ethical guidelines, or even aesthetic expressions is, of course, no easy matter; nor is it simple to politically implement such visions even if we have clarity on such concerns. This is why I refer to naturalism as an aspiration or dream as opposed to a realized reality. Environmentalists orient their policies and actions toward nature. They trust the purity and goodness of nature, and want people to build societies that are committed to preserving nature's wildness and organizing institutions to advance nature protection. The key is the orientation. Realizing the dream may be impossible, but holding on to the aspiration is essential. Environmentalists have certainly had their share of campaign failures and disappointments. The dream of naturalism insists that the goal, however, is still worthwhile. Nature—as the true, good, right, and beautiful—is not something to give up on just because one loses a few battles. The longer and broader campaign still goes on. Engaging in it and deepening one's commitment is to participate in the dream of naturalism.

"Nikola Tesla's Open Dream," Niagara Falls, New York

4

The Dream of Mastery

Hubris was a quaint notion that made for some good plays.
—Simon Young, *Designer Evolution*

In the 1960s, when the back-to-the-landers were reinhabiting farms and trying to align themselves with the rhythms of nature, millions more were giving up rural life for the excitement, economic promise, and cultural attraction of the city. To them, rural life meant a dismal existence in which one was tied to the land and subject to nature's imperatives in an immediate sense. Rural livelihoods depended on winning resources from the earth; one's days were measured by the amount of effort one exerted wrestling with the land. The new urbanites wanted none of this. Cities supplied a different kind of life. Here, one could live many steps removed from nature, free from its dictates and able to experience the many offerings of an urban environment.

In the 1960s and 1970s, much of the world moved from the countryside to the city, and the trend has continued. Today, more than half of humanity lives in cities. This is greater than the entire world population of 1960. To be sure, migration to cities is about more than evading nature. It also has to do with escaping rural poverty, the bonds of traditional family life, and the perceived narrow culture of many rural communities. But for many even these reasons are connected with gaining

distance from the natural world. The crowds, density of buildings, diversity of people, cultural intensity, and varied artistic
expression of the city turns on the ability of people to gain control over and thus minimize the forces of nature in their lives.
To many this is a high in itself. It suggests that the good life—
and as we will see, the moral, aesthetic, and safer life—can be
found not in the woods or on the farm but rather in the concrete, made world of humanity. While the back-to-the-landers
saw nature as the true, good, right, and beautiful, a larger set
of people saw and continue to see these things in humans. The
human being, not nature, is the touchstone of life.

Those leaving the countryside in the 1960s and 1970s obviously did not invent the valorization of the human in contrast
to nature. People have been living in cities for millennia, and
the idea that humanity holds the key to a prosperous, secure,
and worthwhile life has always been a central thread throughout human thought and experience. In Plato's dialogue *Pha-
edrus*, for instance, a fellow Athenian leads Socrates to a grove
of trees just beyond the city's gates. Socrates tells his interlocu-
tor that he rarely visits such places because nature is not his
teacher. Socrates explains that he values only the knowledge of
people in cities.[1] Since at least Socrates's time, many have felt
that the natural world is at best irrelevant to human flourishing
and at worst an impediment to it. Snow, rain, drought, wild animals, and the like threaten human well-being, and protecting
ourselves from nature's elements is a prerequisite to experiencing higher levels of sociality and individual freedom. Nature,
then, is not a model for living, realm of moral practice, or even
place of beauty. Humanity offers these things. In fact, for many,
being fully human requires us to rise above that natural world,
to free ourselves from its bonds. Nature is not, as environmentalism would suggest, the true, good, moral, and beautiful; humanity is.

The privileging of humanity in contrast to nature captures
the central impulse behind many who are skeptical about en-

vironmental worries. Ever since the modern environmental movement arose, critics have argued that environmentalists fail to appreciate the intelligence, innovative spirit, and technical capability of human beings. Critics see environmentalists as fear-mongering, risk-averse naysayers who believe that every act of human control over nature spells doom for humanity or otherwise robs human beings of their dignity. Our genuine freedom, according to critics, indeed rests on such command. Rising above natural constraints offers a world in which humans can satisfy fundamental needs, satiate diverse desires, and exercise the sense of drive embedded in the human spirit. In contrast to naturalism, this sensibility can be called the dream of mastery. It is the aspiration to render nature subject to our vision, will, and desire.

What is the gravitational pull toward the human rather than nature? What is the dream of mastery? Why would we want to control, tame, or otherwise subject nature to our wills? How does the dream of mastery operate in the thought and practice of environmental skeptics? Despite American environmentalism's impressive history and the ancient roots of the dream of naturalism, the dream of mastery has almost always had the upper hand. Indeed, environmentalism has been fighting an uphill battle in a long-standing counter hegemonic effort. To understand this effort and appreciate what is at stake, it is useful to come to terms with the urge to mastery as it is expressed in environmental politics. Moreover, appreciating the aspiration will enable us to understand better the consequences of separating humans and nature in a political context. Together, these will allow us to make sense of the end of nature arguments and how environmentalism might respond.

The True: Prudence and the Promise of Humanity

Environmentalists believe that we should respect nature because it constitutes the biophysical foundation of life on earth, including human life. Environmentalists constantly remind us that we are fundamentally animals, and as such depend on clean air, healthy soil, unpolluted water, and so forth to survive and flourish. Nature, put differently, sets the limits for human life. We must live within nature's bounds *or else*. This is why environmentalists counsel operating in harmony with nature's biophysical character. Our very well-being (and that of other creatures) depends on it.

Critics of environmentalism see things differently. While environmentalists assume that humanity is fundamentally subject to nature—that we are vulnerable to nature's imperatives and vicissitudes—critics believe that it is the other way around. Yes, nature presents limits to human life. We cannot dump indefinite amounts of toxic waste into our rivers, consume untold quantities of oil, pump increasing levels of chlorofluorocarbons into the atmosphere, or otherwise imprint ourselves on the natural world without consequences. But such consequences are not set in stone. Nature may be stubborn but it is not inviolable. In fact, many of the limits that people in the past had identified as nonnegotiable have been transgressed, and often to the benefit of humanity. Human ingenuity, technological prowess, and social engineering have enabled us to stretch nature, push against its seeming thresholds, and ultimately shift its boundaries into the background of our lives or obliterate them altogether. We do this not simply by altering the biophysical qualities of nature—through, for instance, learning how to farm in depleted soil or figuring out how to purify polluted water—but also by adapting ourselves as individuals and societies to changed natural conditions. For example, greater air pollution throughout the world has led to increased cases of asthma and respiratory diseases in general. The study of medicine, however, has provided various drugs that enable us to live comfortably with such assaults on our bodies and thus ignore the so-called constraints that the environment supposedly sets. Nature is not sovereign; humans are.

One sees this orientation in discussions about resources. Environmentalists understand that natural resources regenerate themselves, but only at certain rates, and worry about humans depleting stocks. Environmental skeptics look on such concern as naive or misinformed. The whole idea of running out of resources is a misnomer; resources are, for all intents and purposes, inexhaustible. One reason for this is simply that the earth has so many resources. Environmentalists look at the planet as a small, fragile, finite place that can support only so many people and satisfy only so much demand. According to some skeptics, called cornucopians, this is simply wrong. The earth is a gigantic entity that has untold stocks of resources that have yet to be tapped, to say nothing of being overharvested. To be sure, it may not be economically viable to search for and harvest all of the earth's resources, but this doesn't mean that the resources don't exist.[2]

In this regard, cornucopians constantly question the data that environmentalists offer when it comes to resource depletion. They claim that environmentalists often exaggerate estimates of scarcity or falsely interpret data. For example, in contrast to Worldwatch Institute's annual *State of the World* reports, which track resource depletion and scarcity, the Cato Institute, another think tank, has published *The True State of the Planet*, in which the authors scrutinize data cited by environmentalists and present their own figures on resource trends.[3] In almost every case, the authors show that resources are ever more abundant as humans design more sophisticated ways of locating and extracting natural resources. Likewise, authors such as Bjørn Lomborg and Gregg Easterbrook take issue with environmentalist data on, for instance, the amount of forest cover, biological diversity, and fresh water. In contrast to resource scarcity, they see resource abundance. For people like this, humanity is not about to run out of resources anytime soon.[4]

A second reason that resources are inexhaustible is because market mechanisms will never allow people to extinguish completely valuable raw materials. As economists Harold Barnett and Chandler Morse argued decades ago, as resources become scarce, people will bid up prices, and this will inspire innovation in seeking new supplies or the development of substitutes.[5] The idea is simply that markets will calibrate supply and demand, and that prices will work to check against resource depletion. Higher prices will encourage human ingenuity as entrepreneurs seek to capitalize on increased demand. One sees this today as the price of tuna has gone up so high that people have turned to eating other types of fish. The same thing has happened over the years with the price of natural gas. As demand has risen, so has the price. This has led historically to further exploration and new finds. Some environmental skeptics point out that market mechanisms are so effective at directing people to search for new supplies and develop substitutes that they predict that prices for all resources once believed to be scarce will eventually fall as human ingenuity and entrepreneurship kick in. This found empirical justification in a famous bet between biologist Paul Ehrlich and economist Julian Simon concerning the price of five commodity metals over time. Ehrlich predicted prices would go up as demand increased and scarcity set it; Simon predicted the opposite. In 1990, Ehrlich sent Simon a check for $576.07 to settle the bet in Simon's favor. Lower prices indicate that resources are becoming more readily available or that demand for them has gone down. In either case, resource depletion is avoided.

As mentioned, in the late eighteenth century, Malthus worried about resource scarcity when he predicted that unchecked population growth would outstrip food production, leading to mass poverty, conflict, and human misery. Ever since then, many have warned against overpopulation and its impact on food and other natural resources. Environmental skeptics, subscribing to the dream of mastery, believe this worry is misplaced. Increased population is not a drain on resources but, to the contrary, a contribution to them. While environmentalists look at population growth as a matter of more mouths to feed, skeptics see it as source of greater human intelligence. The next child born may be the next Albert Einstein, and we should never underestimate how much such a child may advance human welfare. Population increase is thus not a matter of quantity but rather quality. Ingenious people contribute to society. Indeed, ingenuity is so fundamental to humanity that the point about Einstein may itself be overblown. Almost *all* children will become working adults, and contribute labor, ideas, and services to humanity. According to Simon, this guarantees a steady path toward a greater quality and even quantity of life for everyone.[6]

This same kind of argument is put forth by those who celebrate not simply increased population but even population density. Agricultural economist Ester Boserup is well-known for demonstrating that population density spurs agricultural intensification, which creates more efficient and bountiful crop production.[7] Various thinkers extrapolate from this, and contend that population is not an excessive draw on resources per se and hence a detriment to economic development but rather a means to more efficient and innovative use of resources. For skeptics, population growth, far from being a hazard to resource utilization, is actually an asset. Simon tries to make this clear when he maintains that the "ultimate resource" consists not of fish, timber, oil, or any other raw material but rather people. Smart, capable, ingenious human beings will find ways of outwitting, adapting, or circumventing nature. Running out of resources will never be a problem.[8]

The trust that skeptics place in human ingenuity extends beyond the issue of resources. In the same way that environmentalists misunderstand scarcity, skeptics claim that they fail to appreciate how human beings can avoid mucking up sinks. Whenever we use a resource—for immediate consumption or as part of the production process—we generate waste. The ecological challenge is to dispose of this in a way that will not harm the earth and, by extension, endanger human beings. While environmentalists see this as an enormous challenge and one that we continue to fail at, skeptics believe the issue is overblown.

One reason for this is connected to the cornucopian orientation that sees the earth's absorptive capacity as so vast that it will never be overwhelmed. Humanity's primary form of waste disposal is dispersion—into the water, air, or soil. The idea is that this will dilute and eventually neutralize harmful substances. Many skeptics count on the vastness of the earth to dilute waste and thus tend not to worry about our waste stream. This is what drives some skeptics of climate change. They believe that the oceans can soak up indefinite amounts of carbon dioxide and that the atmosphere has the ability to wash out much of the buildup of greenhouse gases. Such skeptics go even further in proclaiming that even if certain sinks do fill up (as the oceans have with regard to carbon dioxide), the ecoservice mechanisms of the earth are so enormous and interconnected that countervailing forces will kick in, and that these will compensate. So, for instance, now that carbon sinks have filled and temperatures are rising, skeptics foresee greater evaporation, which will lead to increased cloud cover and hence a blocking of the sun's rays—therewith counteracting further temperature increases.[9]

A second response that skeptics proffer to worries about sinks is simply the generic one about technological optimism. As mentioned, skeptics have much faith in the ability of humans to think themselves out of various challenges. In fact, some skeptics welcome challenges, environmental and otherwise, because they believe that these prod people to be innovative, and therefore to invent or otherwise offer ways to enhance human well-being.[10] Concerns about sinks, then, are not unaddressable or environmental death knells but rather opportunities to further enhance human experience.

There are, of course, some skeptics who recognize the biophysical limits of sinks yet nonetheless worry little about them. These skeptics see that waste is not intentionally produced but instead a by-product of certain activities, and believe that the benefits of these activities far outweigh the costs imposed by the waste. For example, Rachel Carson warned against the overuse of pesticides, including DDT, because these substances fail to wash out easily in the ecosystem, and bioaccumulate in the tissue of plants, animals, and eventually humans. While some skeptics deny that there are any ill effects to pesticides, others in fact do acknowledge this, but feel that the benefits pesticides offer—in the form of, say, greater food production or protection against certain diseases—overshadow the problems. For instance, many skeptics lament the banning of DDT in the United States and elsewhere because they believe that despite the problems DDT poses to ecosystem health, it effectively kills mosquitoes that carry malaria, and the human health gains of DDT use outweigh any other costs.[11] Skeptics make similar arguments about other processes that produce waste including nuclear energy and the production of chlorofluorocarbons.

Skeptics put forth a related argument when they call for ignoring or postponing action to reduce certain forms of waste. Any effort to address waste stream problems requires financial costs—whether this involves changing production processes or treating waste once it is produced. Skeptics often question the wisdom of paying these costs too quickly based on what economists call discounting the future. Because a dollar today will be worth more in the future—if it is properly invested—skeptics frequently propose postponing action on, say, climate change or the permanent disposal of nuclear waste because they believe that future generations will have more financial resources to deal with such issues. Put differently, if present generations can avoid spending money today on certain waste stream problems, they will empower future generations by leaving them the best tool to address any kind of challenge—namely, economic capacity. The logic of this orientation is that it makes little sense to spend today on problems, since doing so would compromise one's present quality of life and reduce the ability of future generations to bring financial power to bear on such issues. This, matched with the faith that humans will continue to perfect their technological capabilities, suggests that the costs of addressing waste stream (and other) problems will ultimately be less onerous overall. At the heart of discounting the future is the belief that economic growth is the supreme tool for enhancing human life. Anything that takes away from this can be seen as an unwise impediment to human betterment.[12]

Whether worrying about resources or sinks, environmentalists have long been labeled doomsayers. They constantly warn that humanity courts danger when it messes too much with nature. The dream of naturalism thus counsels living in harmony with the natural world or suffering the consequences. Environmental skeptics, mesmerized by the dream of mastery, see it differently. Living according to nature is not a prescription for human well-being but for misery. As futurist Simon Young puts it, "Nature deals a cruel hand. Rely on nature and what do we get? Disease, disadvantage, and death, the products of her cold, callous lottery."[13] He asks, does "nature know best when she destroys the lives of millions through disease, earthquake, famine, or flood?"[14]

For skeptics, human ingenuity and technological prowess have always advanced human welfare by beating back, controlling, and ultimately overcoming nature. To them, doom and gloom lies not in transgressing the natural world or otherwise imposing ourselves on it but instead in letting nature rule our lives. Critics contend that valorizing nature over humanity will lead not to a new Eden but rather to greater hardship, starvation, and sorrow. Trying to beat back emotional reactions to Carson's *Silent Spring*, Robert Stevens, the spokesperson for the chemical manufacturer American Cyanamid, wrote, "The real threat, then, to survival of man is not chemical but biological, in the shape of hordes of insects that can denude our forests, sweep over our crop lands, ravage our food supply and leave in their wake a train of destitution and hunger."[15] To put the matter differently, if you want to take humanity back in time to when people scrounged around for scarce food, were vulnerable to widespread disease, and were defenseless against nature's power, then start worshipping and respecting nature. Nothing offers a faster route. The true path to human well-being is through the portal of human inventiveness, intelligence, and technological mastery. Agriculture, electricity, medicine, and so forth have allowed us to opt out of being slaves to nature.

So when it comes to the issue of prudence—of risk and human welfare—the skeptics place their trust in humanity not nature. Nature represents a threat not salvation. Critics of environmentalism give powerful support to this view when they argue that public health, nutrition, and overall human well-being have been improving over the years, and show no sign of reversing themselves due to our increasing mastery over nature. They highlight the fact that life expectancy has been increasing almost everywhere, water-borne diseases are increasingly becoming a thing of the past (at least in developed countries), and food production at least until recently has been growing rather than shrinking. If we fail to respect and propagate human ingenuity and power, we court disaster. Prudence requires us to embrace humanity, and continue to rise above and master nature.

The Good: Humanity as Model for Living

Life is about more than saving our skin. We don't want to merely survive; we want to flourish—to live as fully as possible. Socrates made this point centuries ago when he famously said that the unexamined life is not worth living. Many of us sense that there are some ways of living that are more satisfying, fulfilling, and meaningful than others. Indeed, so much of human culture has been the search for and explication of touchstones for more realized living.

Environmentalists have consistently turned to nature for those touchstones. The dream of naturalism suggests that nature offers a model for how we can flourish. Environmental skeptics, subscribing to the dream of mastery, see things differently. They have no a priori prohibition against examining nature for insight, but question what one finds when one does so. Instead of seeing a world of harmony, goodness, and integrity, they point out that one sees a realm that is brutish, nasty, full of constraint, and just plain unbecoming of civilized human beings. In nature, the cat tortures the mouse; the praying mantis devours its mate while copulating; and hurricanes, tornadoes, and tsunamis ravage ecosystems. Nature, "red in tooth and claw," is the epitome of savagery. Its nastiness is ubiquitous. Furthermore, nature has built into it all the elements of misery. Living beings are born, but they eventually suffer from disease and decay, and ultimately die. To many this doesn't represent something to emulate but rather something to resist and, if possible, overcome. The nastiness of nature was put particularly well by philosopher Mill a few centuries ago when he reflected on how nature treats human beings. He wrote, "Nature impales men, breaks them as if on the wheel, casts them to be devoured by wild beasts, burns them to death, crushes them with stones like the first Christian martyr, starves them with hunger, freezes them with cold, poisons them by the quick or slow venom of her exhalations, and has hundreds of other hideous deaths in reserve."[16] In short, nature is not some obvious realm of goodness and thus deserving of our emulation but quite malevolent. If we mimic nature's ways, we will live as brutes not humans.

The problem with nature is not simply its viciousness but also its restrictiveness. One of the things humans value most is their freedom—our ability to make choices about our lives. Political freedom entitles people to live as they see fit; indeed, we often think that regimes that limit freedom rob people of their ability to experience the fullness of life. For many, nature has always been one of the greatest constraints on human freedom. The vicissitudes of nature—wind, cold, rain, drought, and so on—cramp our style; they restrict, rather than expand, our choices. Nature forces us to be slaves to its imperatives, and this robs us of our autonomy. Modernity took seriously the constraining quality of nature, and recognized that to advance human freedom, we need to subdue nature—to discipline its vicissitudes and master its dynamics. In the words of seventeenth-century philosopher Francis Bacon, nature is something to be "hounded in her wanderings," "bound into service," and "put in constraint."[17] The modern project has been extremely successful in this endeavor, and for many, this success has enhanced the quality of life. Environmental skeptics see humans best able to flourish when they capitalize on the modern project and free themselves from nature's many constraints.

According to critics, when environmentalists look to nature for guidance, they commit what is known as the naturalist fallacy. This is the mistake of drawing ethical conclusions from natural facts. One cannot look at the world, notice empirical phenomena, and then claim that these provide normative vision. This would confuse the difference between "what is" from "what ought to be." When environmentalists see nature as a teacher for human life, they see nature as the good. According to many, this is the epitome of the naturalist fallacy.

Many critics of environmentalism commit the opposite, of course, when they claim that nature is inherently bad or at least offers a model for living that would degrade rather than enhance human life. Somewhere in between are those who are agnostic about nature. They avoid the naturalist fallacy by refusing to ascribe any principled quality to nature. They see nature as neither good nor bad but rather something that simply *is*. This orientation, often espoused by certain skeptics, would argue that environmentalists misunderstand things when they call for emulating nature, since nature provides no normative instruction. Without nature, we are asked to look not to the stars, birds, or biophysical rhythms of the earth for insight into how to live but instead to ourselves. Humanity possesses unique qualities or enjoys a specific status in the world that enables us to draw on ourselves to understand how best to live.

Among those unique qualities, reason plays a special role. While many allege that certain animals are rational beings, humans seem particularly adept at reflecting on and analyzing the world and themselves, or put differently, engaging in abstract thought. Philosophers and cognitive scientists differ about what fundamentally constitutes reason, but few question that rationality is, if not unique to, at least most fully developed in human beings.

Reason allows us to unpack the world. It enables us to examine, scrutinize, and otherwise understand things. When we use it well, we disentangle complexities, dispel mysteries, and gain power as we figure out how best to think and act. To the degree that we accord ourselves with nature, as environmentalists counsel, we use our reason merely to interpret nature's laws and dynamics. In contrast, when we give ourselves the task of rising above nature—of controlling, mastering, or outwitting the natural world—we more fully utilize our reason to the degree that we dissect, manipulate, and ultimately outsmart nature. Reason, in this sense, appears able to click into the *logos* or fundamental "way" of the world that extends even beyond nature. Put differently, it is as if human reason has an expansive reach that can penetrate the nature of *all* things—not simply the principles of the natural world. Thus, we can use it to make more informed decisions about our lives and society, and alter the natural world so we can make good on those decisions.

Humans have always had a complicated relationship with reason. Speaking in broad generalities, in premodern times most people subordinated reason to faith. Living in highly religious societies, people were largely discouraged from seeking evidence for their beliefs and investigating the natural world outside a divine framework. The most important "fact" was that a god or gods ran the world, and our job was not to scrutinize or outsmart divinity's work but rather to celebrate, respect, revere, and abide by it. One sees this, for instance, in medieval medicine, which assumed that illness arises out of sin and that genuine healing can only be facilitated by physicians working in concert with clergy, since healing must ultimately come from on high.[18]

The Enlightenment has long been considered a break with previous times insofar as many societies began collectively to prize reason and allow it to be exercised independent of faith. This resulted in a fundamentally different view of the physical world in which people started to see nature in less purposeful terms—giving up a teleological view of things (in which things have an ultimate purpose toward which they gravitate)—and in more mechanistic ones. Understanding nature in purely physical terms enabled people to drop a deferential attitude toward the natural world, and ushered in an age of unheralded intellectual and technological advancement. Modernity, as such, brought with it miraculous improvements to the everyday lot of people as modern medicine, engineering, and the mechanical arts in general enabled people to free themselves from the threats and inconveniences of nature. One of the most impressive signs of success of modernity is the sheer irrelevance nature has in most of our lives.

Critics of environmentalism celebrate modernity. They see humanity's urge to master nature as a good trait, and one that has brought enormous benefits to our species. They believe that reason, if given free reign, can analyze, disassemble, and redesign nature for human gain.

The dream of mastery maintains that contrary to the environmentalist dictum that nature knows best, nature knows nothing at all. Political theorist John Meyer labels this orientation "dualistic," as opposed to derivative, insofar as it denotes a predominant sensibility in which humanity tears itself away from the cycles of nature by celebrating and elevating qualities believed to be distinctly human. He calls it dualistic because it denotes an unbridgeable chasm between conceptions of nature, on the one hand, and notions of humanity and community, on the other.[19] Environmental skeptics celebrate this gulf, seeing it as what distinguishes us from the mindless, unending rhythms of nature, and the brutishness of the animal and plant worlds.

The Right: Morality's Humanity

Environmentalists look to nature not simply as a set of resources and sinks on which we depend for survival, nor merely as a model to emulate to live worthwhile lives. They also see nature as an object of moral consideration. Nature deserves to be treated with moral respect. Environmentalists differ on the source of this respect. Some see it deriving from the sentience of animals, the dignity of thousand-year-old plants, the magnificence of a given landscape, or even the mere fact that a nonhuman entity exists. Whatever the origins of this moral sensibility, there is a widespread understanding among environmentalists that there is something ethically wrong with treating the natural world however way we wish. Given this, protecting, trusting, loving, or otherwise caring about nature is not simply a pragmatic act or lifestyle choice. It is a moral imperative.

Critics, predictably, disagree. They believe that our treatment of nature is at best an amoral issue. Critics tend to be anthropocentric. They feel that as human beings, we have no choice but to privilege what is important to us and consider our lives the center of all our considerations. Any other species would do the same. As a consequence, they see no imperative to extend moral concern beyond the human realm and believe, more basically, that morality is only applicable among humans. That is, to the degree that we have moral obligations, these are relevant only between people—not between ourselves and the nonhuman world. When environmentalists say that we have a moral responsibility to treat nature with respect, they almost always invoke some form of biocentrism or ecocentrism, which claims that nature has intrinsic value, independent of how humans feel about it. To critics, this is basically nonsense. Things assume value because we endow them with such. It simply cannot be any other way.

Skeptics find biocentrism and ecocentrism amusing but fundamentally misguided.[20] They point out, first, that the choice to be biocentric or ecocentric is itself anthropocentrically motivated. That is, when we choose to care about plants, animals, certain landscapes, and so forth, we do so because it is important to *us*. This is the case even when we do not immediately experience such entities or places. When we value, for example, whales in the middle of the ocean, distant forests, or inaccessible mountain ranges, we do so out of a sense that *we* gain pleasure by knowing that these creatures or landscapes are there even if not experienced directly by people. This describes the feeling that many have when they work to protect, say, the Arctic National Wildlife Refuge (ANWR). The vast majority of people involved in such campaigns will never visit ANWR but nonetheless feel it is important to protect. The environmental skeptic would say that such motivations are not biocentric or ecocentric but rather anthropocentric. *People* get pleasure out of protecting the nonhuman world. There is nothing specifically moral about this. To the degree that people design moral explanations to account for their commitment to nature says more about their need to justify their pleasures than the cultivation of genuine moral probity.

A second and more important critique launched at biocentric or ecocentric ethical thinking is that these are at odds with the Western moral tradition, and thus have no philosophical ground on which to stand. The Western ethical tradition is made up of three main schools of thought, and each has its own reasons for excluding nature from moral consideration. The first emphasizes that we have a duty to act morally toward others independent of the consequences of our actions. Part of this duty stems from recognizing that other people inherently deserve to be treated with respect because of the unique capacities humans possess. Humans enjoy certain cognitive gifts that allow them to reflect on and make decisions about their lives. This rational dimension provides the experience of individual autonomy, and such autonomy is essential to deserving moral treatment. Few of us feel compelled to treat, say, cars in an ethical manner because few of us believe that cars can make decisions or otherwise reflect on themselves. Humans *can* do so, however, and it is this that demands we treat each other with respect. For ethical philosophers like Immanuel Kant, W. D. Ross, and others associated with a duty-based ethics, this means that we cannot treat others as means to an end—purely as instruments of our own desires—but as ends in themselves, for this respects human autonomy. When it comes to plants, animals, minerals, and so forth, most of us deny that these possess the kind of cognitive and willful capacities that enable people to be free, and thus we do not have the same level of moral obligation to them. Plants, animals, and rivers behave according to necessity. They do not deliberate rationally. This denies them moral worth.

A second school of ethical thought focuses not on a sense of duty but rather the results of our choices. Often called consequentialist ethics, it evaluates ethical action according to a calculus of how best to increase pleasure and decrease pain. This brand of ethics, frequently associated with utilitarianism, defines genuine moral action as that which produces the greatest good for the greatest number. This involves a critique of biocentrism and ecocentrism to the degree that most people see humans as most profoundly able to experience the sensations of pain and pleasure. The seventeenth-century philosopher Descartes helped launch this anthropocentric view of sense awareness by declaring that nonhuman animals, plants, and so forth are mere automatons—mechanistic entities devoid of souls and thus insensible to pain. Nonhuman organisms react to stimuli, but they do so without any sense of consciousness. (Descartes accounts for the cries of injured or tortured animals as merely the squeaking of gears.) To the degree that the natural world lacks sense awareness, it is undeserving of moral consideration. Our moral calculus needs to be restricted to the pleasures and pains of human beings. (It should be noted that some use consequentialist ethics to support animal rights and biocentrism. As mentioned in the previous chapter, Singer, working within a utilitarian tradition, ascribes sentience to most types of animals.)

A third orientation, virtue ethics, offers a related critique. Virtue ethics sees moral action not as a matter of duty or mindful calculation about consequences but instead as a practice aimed at cultivating a righteous life. Most often associated with the Aristotelian tradition, virtue ethics is concerned with human flourishing. It believes that we can live in the world and experience our lives in ways that fulfill our deepest potential. The route to this is virtuous action. We should act with, for instance, courage, temperance, honesty, fairness, and magnanimity. These both benefit other people, and develop a state of character in ourselves that leads to moral probity and genuine happiness. Virtue ethics is anthropocentric in that it understands that only human beings are capable of cultivating moral lives. We are able to influence, as Aristotle puts it, our own souls, and flourish best when we do so in a way that implies or expresses a rational principle.[21] Plants, animals, minerals, and so on enjoy no such power over their existence. This does not mean that they are merely mechanistic, in a Cartesian sense—the natural world may actually have its own way of unfolding, or what Aristotle calls teleology—but they are unable to consciously steer themselves in the pursuit of this.

Most people are unaware of the distinction between duty-based, consequentialist, and virtue ethics. Moreover, most are unconcerned with the academic enterprises of professional ethicists. These schools of thought nonetheless implicitly inform our popular ideas of morality and have had a strong influence on the widespread dismissal of nature from moral consideration. By being anthropocentric, critics of environmentalism remain loyal to the Western ethical tradition. As such, they are tone-deaf to the moral stirrings of environmentalists. It makes no sense, according to critics, to extend moral respect to the nonhuman world because natural entities lack the requisites to qualify as morally deserving entities. Natural entities lack the intellectual or other rational capacities that underpin human autonomy, the sensory faculties to experience genuine pleasure and pain, and the ability to cultivate a righteous life. Without these, the natural world sits in an amoral universe.

Undeserving of moral worth, nature can be treated as we see fit. This explains and justifies the dream of mastery. Without moral obligations toward nature or a sense that nature has intrinsic worth, we can seek to control it and render it subject to our will with no remorse. In fact, we flourish most when we do so. To be sure, we may choose not to do these things for pragmatic reasons, but this is a far cry from a moral injunction.

The Beautiful: Mastery's Aesthetic

Environmentalists see nature as a fundamental source of beauty. Mountains, orchids, forests, and oceans provide tremendous visual and other sensory delight. As Leopold mentions in the opening of *A Sand County Almanac*, however, not everyone shares this aesthetic judgment. Indeed for many, nature is boring. It moves slowly, lacks an articulate voice through which it can communicate depths of meaning, and appears bland compared to the drama and excitement of human society. For this reason, while many people do not necessarily denigrate nature or dismiss the aesthetic judgment of environmentalists, they nevertheless feel that the natural world is simply not for them. It does not resonate with their aesthetic sensibility and hence holds no special status in the pantheon of what counts as genuinely beautiful.

Aesthetic judgment is highly subjective. It changes across time and culture, and varies among individuals. It therefore should come as no surprise that people differ over what appears as aesthetically pleasing. But there is something deeper at work here. It is not just a matter of chocolate versus vanilla, or Paul Cézanne versus Jackson Pollock. There is a difference about the source of beauty. For most of history, many thinkers of aesthetics have seen nature as the supreme embodiment of beauty and identified quality art as that which best imitates the natural world. Beauty is a matter of color, light, tone, volume, and form, oriented around principles such as harmony, proportion, and unity, and these elements have seemed, to many, to come together most elegantly in the natural world. The reasoning behind this is complicated. It may have to do with the giveness of nature—the sense that it comes from God or the universe itself, and thus contains the most fundamental and universalist dimensions of beauty. Alternatively, it may have to do simply with the immediacy of nature, and the potent quality of its forms, color, and so forth. For our purposes, accounting for this preference is less important than noticing it. Environmentalists relate to this aesthetic sense. They see nature as the supreme source of beauty in the world. This does not mean that they always prefer art that imitates nature, or that they are blind, deaf, or otherwise turned off to other aesthetic expressions. It simply means that environmentalists feel a deep aesthetic affinity with nature. This inspires them to experience the natural world through hiking, biking, snorkeling, and the like. Exposed to nature, many environmentalists feel that they are undergoing a profound aesthetic experience. At the highest level of abstraction, they are clicking into the meaning of beauty itself.

Critics of environmentalism may see such experiences as worthwhile, but their aesthetic preferences at this most fundamental level lie elsewhere. They see the source of greatest beauty in humanity. Humanity offers the most dazzling, dramatic, and even picturesque elements of aesthetic pleasure. In the same way that critics feel that our safety, sense of life, and morality are best assured by improving on nature, they also believe that beauty involves the same kind of transfiguration and enhancement.

We often associate aesthetics with art. While beauty involves color, light, tone, and so on, and consists in qualities like symmetry and proportion, it is the way the artist arranges, constructs, or presents these elements that makes them beautiful. Art is, after all, a human enterprise. People envision and express themselves through art; art cannot simply be found. (To be sure, artists of various traditions have used found objects as expressions of their aesthetic message, but what makes a rock, say, a piece of art versus a natural object is the degree to which the *artist* arranges, frames, changes, or otherwise interprets the rock's presence.) This sense of made as opposed to found is key to the aesthetic judgment of many and hints at the difference between how environmentalists versus their critics orient themselves toward aesthetic pleasure. For environmentalists, nature provides many of the visual and other sensible elements of aesthetic appeal. It is the connection to nature, the unbidden or given quality of natural things, that infuses them with beauty. For critics, it is almost the opposite. They recognize the made quality of art itself, and thus do not necessarily long for or sense the artistic quality of a piece of work in its gesturing toward or imitation of the natural world. Again, this doesn't mean that environmental skeptics fail to find any aesthetic pleasure in the external, natural world. Rather, it simply suggests that nature unto itself is not their most sublime example of genuine beauty.

This helps explain the difference between those who go to nature for recreation and those who go to the city. Every year, millions of people travel to Paris, London, New York, and Rio de Janeiro for the aesthetic experience of being in a place packed with people, and in which human expression finds some of its most intense representations. This is also why people choose to live in such places. Indeed, the bustle and cultural richness of cities hold this appeal. To people who enjoy such things, the Rocky Mountains may indeed be pretty, yet they are nothing compared to, say, the skyscrapers and commercial districts of Manhattan or Tokyo. In such cities, what some take to be grime, din, and congestion, others appreciate as the face of human beings living together. Furthermore, above the debris and density of the city is what urban theorist Jane Jacobs calls "offerings." These are the abounding cultural riches that often develop in the synergistic avenues of urbanity.[22] Here, beauty emerges from what humans do together on top or independent of the natural world. The idea of "on top of" is crucial in this context. Many see great art as a matter of refined taste. They understand the development of an artistic sense to be something that has to be cultivated, and this suggests an element of civilization to it. It is as if one rises above the immediacies and seemingly primitive quality of nature per se, and nurtures (or, one might say, cultures) a less brutish, more sophisticated sensibility. Now this may sound like high culture as opposed merely to aesthetic pleasure, but there is a connection. The attraction to things human finds the true, good, moral, and now beautiful in the artifice of the world, in the human built environment not the natural one. This may be why we call someone of heightened artistic sense "urbane." More generally, it suggests that one need not go to the woods for visual or other sensible delight but rather to the museum, marketplace, or society in general.

At a higher level of abstraction, and one that may appear as an intellectual stretch but still relevant, the aesthetic appeal of the human-made world in contrast to the natural one rests on a long-standing philosophical preference that values the life of the mind or soul over that of the body and physical experience. For millennia, theologians and philosophers have denigrated bodily life because it is impermanent and somewhat impure compared to other realms one could imagine. Material objects change over time, decay, and if they are living beings, ultimately die and pass away. For many, such a realm does not deserve our deepest commitments because it will undoubtedly disappoint us, and more important, it must be less real than dimensions of the universe that seem to be eternal.

Physical objects come and go. Aristotle described their trajectory through time as a matter of "becoming"—an entity grows into its own nature. So an acorn germinates and develops into an oak tree. Someday, the oak tree will die and return to the soil from which it emerged. The material world is like this more generally. It is always undergoing change. Behind the dramas of becoming, however, numerous thinkers have conceptualized dimensions of the universe that do not change. In the same way that a person remains a singular entity through birth, adolescence, adulthood, and old age, thinkers have imagined some basic forces operating in the universe that fundamentally stand behind or otherwise secure the phenomenal world. In contrast to becoming, this element is a matter of "being," and represents the unchanging qualities of things such as the "isness" of reality or the seemingly immaterial dimension of consciousness. Philosophers and theologians have long debated the relationship between being and becoming, and have wrestled with questions about privileging one over the other. Through these debates, though, there has always been a strong preference for seeing the suprasensible world as somehow superior to the material one.

Theistic religions use the concept of God or gods to draw this distinction, and make clear this preference. God is a being that possesses some combination of omnipresence, omniscience, and omnipotence, and operates in the universe from behind the scenes—in some nonmaterial realm that fundamentally animates the material one. God, we are told, created the world and thus stands beyond it, even if for some God's presence pervades it. Many nontheistic thinkers equally privilege such a realm by imagining it to be a dimension of consciousness or abstract thought itself that may not pull the strings of the physical world but nevertheless exists "on high" as a domain that enjoys a higher order to it. One sees the privileging of this in those who understand the physical realm as representing a corrupted imitation of the suprasensible one. Such a view was put most clearly by Plato, who imagined the suprasensible world made up of ideal forms that can be grasped only by leaving sensate experience and ascending toward a theoretic reality. Bodily entities represent instances of this realm, or put differently, participate in this realm to the degree that they approximate, in the crude materiality of mineral, fiber, and muscle, the pure ideal that is beyond materiality altogether. In this sense, the phenomenal world will always be one notch below—and hence represent an imperfect rendition of—the suprasensible one. This is why Socrates believed that the philosopher was always dying since she or he ultimately sought to leave the bodily world and dwell in the highest of realms.[23]

Those resonating with this long tradition—which obviously started before and has certainly not disappeared since Plato—have always seen the ultimate meaning of human life connected to contemplating or otherwise approaching the supramaterial world, and it is this orientation that supports the aesthetic sense associated with the dream of mastery. Nature is frequently associated with the body; it is devoid of mind or soul, and operates unreflectively. It is, in other words, the epitome of physicality. It consists of objects that can be touched, heard, smelled, or seen—not ideas that can be imagined, dreamed, or contemplated. As such, it could never serve as the most important source of aesthetic beauty. To be sure, the made world of humans is also empirical, can be sensed, and will change over time; it thus may seem no different from the natural one. Yet the difference is that human-made, artistic expression is like an arrow pointing to the suprasensible realm. Its depictions of even landscapes or other seemingly natural places and abstractions stir an emotional, intellectual, or even spiritual response of an abstract quality. It is as if artifice is an avenue of departure from nature toward suprasensible reflection. It is in this sense that the dream of mastery finds its highest aesthetic pleasure in the human-made, artistic world.

Conclusion

The impulse to conquer nature and render it subject to human design runs long and deep through history. Many critics of environmentalism resonate with this impulse or what I call the dream of mastery. They belittle nature, and thus see no hesitation in beating nature into service or otherwise subjecting it to human desire. Critics believe that the true, good, right, and beautiful can be found not in nature but instead in humans. They would say that the dream of naturalism, which animates so much of the environmental movement, is a nightmare. It promises danger, a dismal quality of life, moral confusion, and aesthetic naïveté.

I spent time detailing the dream of mastery in this chapter and the dream of naturalism in the previous one to uncover the philosophical roots of the environmental politics. Environmental politics, at bottom, is an argument between two fundamentally different worldviews. The first, the dream of naturalism, sees human well-being tied to aligning human practices with nature. The second, the dream of mastery, sees the opposite. It associates human flourishing with overcoming the constraints of nature and living in as much freedom as possible. The dream of mastery, in other words, suggests that humans will realize their highest potential as individuals, societies, and species to the degree that we can manipulate the natural world as we see fit. *This* is genuine humanity.

If environmental politics has been polarized around these twin dreams, this is about to change. As I will show in the next chapter, the grounds for both dreams are difficult to justify anymore. The increasing disappearance of the wild, in an empirical sense, and the conceptual end of nature, as evidenced in social constructivist ecocriticism, are fast rendering both dreams anachronistic. Appreciating this is critical to moving beyond the conventional polarization of environmental politics, and developing a meaningful environmentalist stance in a postnature age.

"The Ruins," Lama, New Mexico

5

The Great Vanishing: Into the Postnature World

Man, too, is making many far-reaching changes. The most influential half animal, half angel is rapidly multiplying and spreading, covering the seas and lakes with ships, the land with huts, hotels, cathedrals, and clustered city shops and homes, so that soon, it would seem, we may have to go further than Nansen to find a good sound solitude.

—John Muir, *Our National Parks*

One of the oldest dreams of politics is to rise above it. Politics is about power. It involves determining, as political scientist Harold Lasswell puts it, "who gets what, when, where and how" within a given community, and this is often an ugly business.[1] It is marked by competition, persuasion, and conflict as various groups fight to advance their interests and values. What an amazing thing it would be if we could ascend out of the bickering and horse-trading of politics, and order our collective lives according to principles that everyone would recognize and dicta that could be easily translated into practice. The great dream of politics is to discover a realm outside the political world that would set the standards for human conduct.

In many ways, the dual dreams of naturalism and mastery represent the great aspiration of politics. They aim to identify extrapolitical principles to orient our lives. When environmentalists, for instance, see nature as the true, good, right, and beautiful, they understand it as a privileged source of value above the fray of politics. Nature is, after all, "natural," and not a matter of opinion or choice. It is part of the given rather than the made world, and it is this that makes it worthy of our respect and adherence. As historian Cronon explains, "When we speak of 'the natural way of doing things,' we implicitly suggest that there can be no other way, and that all alternatives, being unnatural, should have no claim on our sympathies. Nature . . . becomes a kind of trump card against which there can be no defense."[2] Environmentalists invoke nature in their political struggles as such a trump card. They turn to it to say, essentially, "We are advocating not merely our opinion about how the world should be but rather what nature itself demands." Nature, as such, becomes a biophysical, instructional, ethical, and aesthetic imperative. To put it differently yet not inaccurately, it represents a type of deity to which environmentalists appeal in their political maneuverings.

Environmental skeptics engage in a similar practice with their valorization of humanity. When they see humanity as the true, good, right, and beautiful, they are suggesting that there is something inherent within the human being—a human nature—that likewise is uncontaminated by mere judgment, opinion, or social context, and therefore deserves respect and conformity. For environmental skeptics, what is most important to notice about the human being is our innate capacity to reason and ingenious spirit. These represent fundamentally what we are about. To be fully human, then, requires giving a free rein to these essential attributes even—and especially—if this enables humanity to unlock nature's secrets and manipulate them. Indeed, far from being a transgression or hubristic excess, such forays into the nonhuman world are nothing more than living according to *our* deepest nature. This devotion to humanity represents an attempt to rise above politics insofar as human nature is offered as the standard or informing principle for political life. Environmental skeptics invoke humanity to say, in essence, "We are advocating not merely some personal or interest group whim but rather policies that reflect the nature of what it means to be human." Humanity, in other words, becomes the uncontested source of prudential, ethical, moral, and aesthetic wisdom and practice. While not a god per se, it represents a deity of sorts in that it sits uncorrupted at the core of our being as a species.

While environmentalists and their critics have long dreamed of leaving the world of politics, their respective gods—nature and humanity—ensure continual political conflict. The two terms have acted like ideational magnets drawing people to one pole or the other, and together have framed environmental disputes in terms of having to choose one center of value over another. Indeed, for too long environmental politics in the United States has been not simply a set of fights over this particular wildlife preserve, toxic waste dump, or even the specifics of climate change. Rather, it has been a philosophical dispute about our place in the cosmos and how we should understand our relationship with the more-than-human world. Our choice has been between nature or humanity—with little middle ground.

It is time to move on. It is time to move on because the defining categories of the debate no longer make sense. While we used to think of humans and nature as self-subsisting realms marked by distinct characteristics and qualities, the two spheres are melding into each other such that it is hard to draw a boundary between them. Empirically, we humans have extended ourselves across and into every ecological niche on the planet, making it impossible to say anymore where humans end and nature begins. Likewise, conceptually, we have come to understand that neither nature nor humanity has a given "nature" to it, since our ideas of each are social constructions. This rids nature and humanity of their theistic qualities, and prevents us from turning to them with confidence for fundamental insight when negotiating our way through environmental issues.

In this chapter, I describe the empirical and conceptual ends of nature. I show how humanity has blended itself with the more-than-human world and explain the way we socially construct nature—both *human* nature and *nature's* nature. My aim is to demonstrate the increasing poverty of the dual dreams of naturalism and mastery. This is important to the overall argument of the book insofar as recognizing that neither pole provides fundamental grounding enables us to embrace a post-nature world and construct an appropriate environmental politics.   .

End of Nature: Empirical

Take a moment and look around. If you're indoors, gaze out a window or walk outside. If you're outdoors, lift your eyes and survey the environment surrounding you. Further, have a sniff; listen. Where is nature in all of this? Where is the natural world?

In chapter 2, I described how American environmentalists have worked for decades and arguably centuries to build barricades between us and nature. They have done so to protect the environmental services that nature provides to human beings and maintain the well-being of the more-than-human world for its own sake. How well has American environmentalism done? Has the natural world been sheltered from overly exploitative human practices? Does it enjoy a life on its own in which it can unfold independent of human will and purpose? Has environmentalism saved the wildness of nature?

As I look out the window of my relatively suburban home, I see houses, telephone wires, pavement, and automobiles. I hear the soft rumble of cars going by and airplanes flying overhead. In this, I have a hard time finding nature. When I do, its character becomes suspect. Take, for example, the magnolia tree I see just outside my back door. While the telephone lines and so forth are not nature, surely this is. It was germinated from seed, grows in the rays of the sun, is nourished by water, and stretches its limbs and leans toward the sky out of an act of instinct or biological mechanics. The plant operates according to, in other words, a telos; it has a way about it that is encoded in its DNA—or, if you prefer, its soul—that animates it. The tree is self-willed; it does its thing separate from the human world.

But as I consider the tree in more detail—reflecting on its placement, health, and overall physiognomy—it appears less natural than at first sight. In 1703, Charles Plumier classified the first specimen of magnolias on the island of Martinique. Over the centuries, various species of magnolias have been classified and new breeds have been developed by horticulturalists through crossbreeding. The magnolia out my window is probably a hybrid species planted a half century or so ago. People most likely watered, fertilized, and certainly trimmed it over its lifetime.

To me, the magnolia is natural yet not completely so. That horticulturalists have cultivated the plant from various specimens, and that someone actually planted the tree in my yard suggests that it is not simply an expression of nature itself nor did it simply arise out of the environment on its own. Furthermore, all the watering, fertilizing, and trimming indicates that the tree has not just unfolded independently from human beings but also has been developing in partnership with humanity. Right now, the tree has dropped a huge number of its leaves. Magnolias do this when they are short on water in an attempt to conserve moisture. This seems natural enough, but when I realize that this area is experiencing one of the worst droughts on record for this time of year and that some observers are linking the drought to anthropogenic climate change, I realize more and more how much humanity is spliced into the tree.

Okay, if the tree isn't fully natural, how about the birds? How about the nuthatches, orioles, cardinals, and sparrows I see whipping across the sky, and landing on the magnolia or my neighbor's roof? It is a tough call. I know that many of my neighbors feed nuthatches suet, a type of cake derived from beef kidney fat, and try to attract orioles by providing sugar water. Simply providing such food has an effect on the range of specific flocks of birds, and the content of the food certainly has an influence on bird physiology. Moreover, the birds are in a neighborhood of lawns, telephone wires, buildings, and lots of concrete. Their lives are shaped accordingly. Finally, the birds, like all living creatures, are sensitive to the amount of moisture, temperature, and storms in the area—things that humans have increasing influence over due to the urban heat island effect and climate change. The birds, for all their beauty and otherness, are flecked with a human signature. We have not manufactured them but we surely play a role in their lives. They are not living free from human will and purpose.

The line between humans and nature may be only slightly transgressed with regard to trees and birds. It is all but obliterated when we consider how deep and wide the human imprint is across the planet as a whole. These days we are not merely influencing the lives of nonhuman creatures or inflecting a human presence on the land but fundamentally revamping the earth's ecosystems. We build cities and cultivate crops on most arable land, draw resources from the depths of the earth, and disseminate waste across the sky, into the earth's waters and soil. More profoundly, we literally move mountains, reroute rivers, empty out seas, and denude whole forests. According to environmental historian John McNeill, the human footprint is so large these days that we penetrate every realm of the earth: the lithosphere, pedosphere, atmosphere, hydrosphere, and biosphere.[3] In the process, we have gotten rid of the distinction between humans and nature.

The Litany: Water, Sky, Land

Take the oceans, for instance. For centuries, the oceans were seen as vast expanses against which humanity seemed puny and was certainly powerless to exploit. Humans could navigate the oceans, and draw fish, whales, and other living creatures from their waters. But for most of history, humans barely skimmed the surface, as it were. Our limited technologies as well as our smaller numbers restricted our impact. This changed in the early part of the nineteenth century. Steam-powered factory ships came on board, followed by the introduction of trawlers, and later, drift, sine, and bag-shaped trawl nets in the twentieth century. These technologies took another qualitative leap a number of decades ago as sonar, underwater video cameras, and geographic information systems became part of the fishing enterprise. These have allowed ocean trawlers to drag miles-long nets across increasingly greater expanses of the oceans. For example, every year trawlers pull their nets through every square foot of the Dutch region of the North Sea.[4] More generally, contemporary fishing has depleted the oceans of fish such that the United Nations has claimed that two-thirds of global fishery stocks have been overfished or fully exploited.[5]

The human presence in the oceans comes not simply from fishing its waters but also from using the oceans as a dump site. Despite international regulations, humans throw millions of tons of paper, plastic, metal, and industrial, radioactive, and household sewage into the oceans each year, and let various chemicals and other waste runoff into the earth's waterways. Researchers have found remnants of this waste in all five oceans and at great depths. According to one observer, "There's not a liter of water anywhere without its share of PCB and DDT."[6] Furthermore, plastic has made its way into almost every ocean, including the northernmost reaches of the Arctic Ocean and the sediment of the ocean floor. It appears that our reach into and across the oceans extends far, wide, and deep. And as our reach grows, the scope of nature's wild otherness shrinks. The oceans are more human every day.

Human reach goes beyond the oceans. It extends greatly into the earth's landmasses and the habitats of all creatures. This has made humanity not simply another species in a vast evolutionary process but instead the actual governors of the process itself. Paleontologists tell us that according to the fossil record, there have been five great extinctions in the geologic past. The last one was roughly sixty-four million years ago, when the dinosaurs disappeared. Conservation biologists maintain that we are now in the midst of the sixth great extinction, and this time we know the cause is not an asteroid or some other cosmological event. Today, humans are destroying diverse habitats, hunting untold species, and otherwise compromising the ability of creatures to live on earth to such an extent that close to a hundred species are disappearing every day, and roughly one-quarter to one-third of all species will become extinct within the next fifty years.[7] This diminishment has already reduced the variety of genetic diversity among species to such an extent that the evolutionary process itself has been greatly curtailed. According to conservation biologist Michael Soule, human action has caused such significant evolutionary change that "the evolution of new species among large vertebrates has come to a screeching halt."[8] Put differently, we have extended our reach so deeply into the plant and animal worlds that we have essentially taken over the reins of evolution. We have become what some call the "governors of evolution."[9]

As is well-known, humans have also changed the skies. The planet possesses a stratospheric ozone layer that protects living entities from harmful ultraviolet radiation. For millennia, the constitution of the ozone layer remained fairly constant. This changed in the 1930s with the invention of chlorofluorocarbons and the increasingly widespread use of other ozone-depleting substances. By the end of the last century, ozone levels had fallen so much that skin cancer and cataract rates in humans had gone up, phytoplankton (which form the backbone of the marine food chain and provide significant amounts of oxygen for the earth) showed signs of reduction, and susceptible animals such as certain species of frogs exhibited signs of deformation or disappeared altogether.[10] Although the 1987 Montreal Protocol and its subsequent amendments have significantly curbed the production of ozone-depleting substances, the ozone layer has yet to mend itself, and there are critical questions as to whether this is even a possibility. Not only is the increased use of hydrochlorofluorocarbon's, the preferred substitute for some of the most aggressive ozone-depleting substances and other gases such as nitrous oxide, compromising the protocol's effectiveness, it is unknown if the ozone layer can ever fully right itself.

These conditions and many others that can be cited are well-known. Most of us are familiar with the litany of environmental problems and the kind of comprehensive impact we have had as we systematically de-wild the earth. I rehearse some of the litany and emphasize humanity's commanding presence here not to provide new details but rather to frame environmentally harmful actions and our extensive presence in such a way that we can appreciate how they are fashioning a new set of challenges for environmentalism. Environmentalists have long worried about the threats that pollution, loss of biological diversity, and ozone depletion pose for humans and other creatures. Additionally, we have been troubled by the increased humanization of the natural world. Traditional environmentalism, however, has always assumed that the nonhuman world forms a benchmark against which we can measure degradation and a domain that can stand for what the earth is like absent human influence. Our steady encroachment and indeed colonization of the air, water, soil, and species is removing that vestige of environmental insight. As mentioned, environmental writer Bill McKibben famously refers to this loss as the end of nature. He claims that we have empirically erased the divide between the human and nonhuman worlds by making so many inroads into nature that we must abandon the whole idea of a self-willed, other-than-human world. We now live on what environmentalists call a "full planet," where humans are essentially everywhere. Reaching this point is a profound historical event, and coming to terms with it offers a great challenge to environmentalism.

The Litany from the Inside Out

The profundity goes deeper, though. As McKibben and others talk about the end of nature, they provide a linear narrative. They explain how humans have increasingly encroached on nature's domain, and how eventually the circumscriptions have became so narrow that there is now no longer anything we can comfortably call natural or wild. Implicit here is that we can one day pull back from encroachment and resurrect wildness. That is, one day nature may regain a foothold and live out its life free of human influence, if we sufficiently withhold our powers of intervention and stop trespassing.[11]

Crowding out nature is only one part of the problem. Recent advances and practices in biotechnology, nanotechnology, pharmacology, and artificial intelligence point to a different type of colonization—one that can never be undone. Here we are not so much circumscribing nature as splicing ourselves into its very processes, and in turn, changing our own human identity and the identity of the so-called natural world. Humans, animals, plants, and machines are now morphing into each other, so much so that the entire idea of a divide between the human and nonhuman—with the quiet hope that this divide may one day be resurrected and expanded—is folly.

In his book *Evolution Isn't What It Used to Be*, political scientist and futurist Walter Truett Anderson chronicles successive augmentations to the human body.[12] From eyeglasses and hearing aids to pacemakers and artificial limbs, technologies have allowed so many of us to become part human and part machine. We see this, for example, in synthetic materials used in medicine that resemble and can ultimately bond with human bone and tissue. We see it also in dialyzers and artificial organs that perform essential biological functions. According to Anderson, these innovations are creating genuine bionic bodies. They are enabling us to engineer ourselves by designing mechanical and electric apparatuses that can be folded into our very bodies.

The correlation to this is the organic machine. Advances in artificial intelligence, robotics, and the like blur the lines between the living and nonliving, inviting artifice into the privileged fold of the organic. Machines now move, self-design, and even learn. To some, this is shifting the balance of animation. As University of California professor Donna Haraway puts it, "Our machines are disturbingly lively, and we frighteningly inert."[13] As we increasingly rely on machines, they take over more of our lives, conditioning us to live according to mechanical rhythms rather than biological ones. The extreme of this is some sort of *Matrix*-like world in which we humans become subject to our own technical creations. Short of this, of course, is simply the situation in which the mechanical increasingly influences the biological so that the latter is no longer a distinct realm. It is contaminated, as it were, by the artificial, and it may be impossible to pull the two apart. (Perhaps one of the more dramatic instances of melding the mechanical and biological is the manufacturing of the entire genome of a bacterium by stitching together its chemical components. Such "advances" represent the beginnings of what could be called "synthetic life.")[14]

On the other side of the divide, humans are increasingly melding our bodies with plants and animals. At the most basic level, humans have long domesticated various creatures, and have made them part of our diets, medicines, and cosmetics. The early use of insulin from cattle and pigs to treat diabetes is perhaps one of the most obvious examples of this. One also sees it in the use of animal organs for human transplants, and in production processes that use animals as "living factories" whose bodies produce proteins, peptides, and modified amino acids to supplement human endocrine systems. In one of the more interesting interspecies medical innovations, scientists are now creating human skin replacements by combining the cells taken from the foreskins of baby boys after circumcision with purified collagen from cows. These types of procedures and advances are erasing the boundary between nonhuman animals and people.

The most extensive types of such erasure can be found in the emerging world of biotechnology. Among its many variants, splicing genes from one species into another to affect certain characteristics represents one of the deepest human cuts into nature. According to physicist James Trefil, our ability to manipulate genes has allowed us to "get under the hood of living systems."[15] Once there, it appears the sky is the limit. So far, we have been able to do relatively simple things like inject vegetables and fruits with salmon genes to avoid freezing, and engineer potatoes to produce their own insecticide. We have also bioengineered a bovine growth hormone that increases the amount of milk a cow can produce by 20 to 30 percent, created mice programmed to grow certain cancers, and inserted genes into corn to kill off corn borer. And we have created cows that can, with the help of an extra synthetic humanlike chromosome, manufacture human proteins in their blood. Along these same lines, we have inserted a human gene that produces the blood protein antithrombin into goat DNA and have been using the offspring involved to produce large amounts of the protein for medical purposes. We are apparently also on the cusp of building "humanized cows" that can have complete human blood in their systems and thus provide a source of plasma for human transfusions. The list of such bioengineered creatures goes on: cows that can produce fortified human milk, rice that can generate vitamin A, bananas that can grow vaccines, and crops that can produce their own pesticides. Such bioengineered creatures represent fundamentally new species. They did not evolve out of the long evolutionary process but rather came into being as a human fabrication. Far from emerging out of the wild such life-forms are being cooked up in laboratories and stamped (and often patented) by the human fingerprint. In this sense, bioengineered species are arguably closer to artifacts than so-called natural entities.

When environmentalists rehearse the litany of environmental woes, they usually do so to warn against our steady encroachment on the natural world and how we are reformatting wildness in our own image. Bioengineering represents the outer reaches of such encroachment. It involves not simply pushing nature to the edges of our experience but rather getting inside of life itself and rewriting its genetic instructions. This, indeed, is the next frontier of our ascension from nature and our entry into not only a postnature world but also a posthuman one. Such a world represents not just another step in humanity's long trajectory of insulating ourselves from nature; but also the rewiring of ourselves as a species. If our ascension from nature has thus far been about freeing ourselves from nature's "givens"—nature's biophysical constraints as we find them—then the end of humans involves freeing ourselves from our genetic givens. It enables us to design human life anew.

There are tremendous debates these days about how much human experience is a matter of nature versus nurture. Those emphasizing nature are increasingly on the defensive as cultural critics point out how so much of what we take to be natural is really a matter of sociocultural constructions (an orientation that I will examine in greater detail in a moment). We are told that our sexuality, psychology, and sociality, for example, are not expressions of deep-seated, given characteristics but simply of predominant discourses or ways of socially ordering our world. And yet few deny that genes play a central role in the unfolding of individual human lives. Genetic codes determine hair color, height, sex, and according to many, predispositions and certain talents. At the outside, our genetic constitution makes us human. It distinguishes us from dogs, cats, orchids, and whales.

Bioengineering offers the possibility of our altering such distinctions. It promises to unleash us from our genetic makeup so we can design ourselves as we see fit. According to many observers, now that we have mapped the human genome, we will soon be able to master the capacity to alter our genetic makeup in ways that will give us the ability to determine who and what kind of beings we want to be. This includes the ability to decide what kind of children we wish to have—or through cloning, if we really want to have children at all or merely replications of our individual selves. It would allow us to design people to have certain physical characteristics, particular talents, and even—if geneticists are right—specific psychological dispositions. We could, for instance, design children to have blue eyes, be predisposed to become a super athlete or master musician, or to feel happy-go-lucky toward life as a whole. Of course, we wouldn't have to stop there. Such choices remain limited to messing around with human genes. Many imagine a future in which we manipulate and import so many genes from other creatures that we start literally sculpting the human genotype. For example, Dartmouth Medical School professor Joseph Rosen once remarked that he would engineer a person with wings if given permission by a medical ethics board.[16]

To many, this trajectory is scary not only because of the mangling of human life that we may inevitably see but also because of the principled concern of cutting the increasingly thin thread that connects us to what it means to be genuinely human. Whatever we have become through cultural innovation, our biochemical makeup and human consciousness itself emerged through the process of evolution. As we become governors over our own evolutionary path, we remove the connection we have to the giveness of evolution and turn it into the made. Some observers see such business as a matter of pseudoextinction in which progressive self-transformation through germline tinkering changes us into a fundamentally different life-form such that we leave Homo sapiens behind.[17] Such a posthuman future would not only represent the end of humanity as we know it but the complete taming of our own wildness as well. Our genetic lot presently involves an element of chance in which the unpredictability of the universe as it unfolds expresses itself. Mastering such unfoldment represents another instance of eradicating wildness as we try completely to control, civilize, or tame the feral dimension of life itself. Such a prospect has led one observer to announce, "We have entered an era when our very humanness, in genetic terms, is no longer a necessary condition of our existence."[18]

Bioengineering and the other practices I have been describing illustrate the empirical end of nature. They all but erase the boundary that we have long thought separates and distinguishes humanity from the wider world of plants, animals, mountain ranges, and oceans. As the divide fades, so do the dreams of naturalism and mastery. As I will show in subsequent chapters, this challenges American environmentalism and yet also opens up promising avenues for forging a postnature environmental politics.

End of Nature: Conceptual

While the end of nature is a historical event, there is another dimension of nature's demise that equally challenges environmentalism and suggests a new course for American environmental politics. This is the social constructivist understanding of nature. Throughout history various thinkers have understood that human beings impose meaning on the world. The categories that we use to make sense of experience and norms that animate our lives are not written into the nature of the universe but are instead cultural "facts" that societies have developed and reproduce in one form or another across generations. They are narratives or discourses.

While certain thinkers and whole schools of thought have entertained a social constructivist understanding through the ages, the general orientation seems to be gaining significant ground these days. Globalization has encouraged the mixing of cultures, and this is enabling people to experience outside viewpoints on their own understandings. Increasingly, this is leading many people to adopt a more relativist stance toward their own beliefs and practices. Some label this era "postmodern," to highlight the abandonment of modernist certainties and an appreciation for contingency. Yet because the sensibility is not a historical phenomenon but rather a perennial insight, it makes more sense to call the orientation social constructivist. Constructivism implies simply that the meaning of our world is not self-revealing but a social projection.

The widespread subscription to social constructivism challenges the dreams of naturalism and mastery. Both naturalism and mastery assume that there is a fixed, essentialist manner to their respective "deities." Many environmentalists, especially those squarely in the American environmentalist tradition, assume that there is something inherent to nature that is beyond social construction, and thus something we should respect and orient our lives around. Skeptics believe the same thing about humanity. There is something intrinsic to Homo sapiens that is beyond culture and sociohistorical context, and hence something that deserves our attention, respect, and adherence. Like the empirical end of nature, a social constructivist outlook questions both of these assumptions.

Take a look around again. Gaze out the window or walk outside. Sniff, listen, and feel the world. In doing so we can ask ourselves once again, "Where is nature? Where is the genuine other-than-human world?" When we did this earlier, we found it hard to locate nature, as we recognized how deeply and extensively the human signature is etched on to the world around us. This time, as we explore the social constructivist critique of nature, we arrive at the same difficulty, only through a different but no less profound route.

Remember the Magnolia tree? It appeared to be not fully natural to the degree that human beings have been responsible for cultivating it as a particular species over the past few centuries, and in the case of my specific tree, in the sense that people have planted, watered, fertilized, or otherwise influenced its life. At a higher level of abstraction, we should also notice that the human imprint comes not only when we consider what has been done to the tree but also the fundamental meaning of the entity itself.

As mentioned, Plumier first classified magnolias on Martinique in 1703. Although the plant was known locally as "talauma," he gave it the genus name, *Magnolia*, after the French botanist Pierre Magnol. It turns out that Plumier's appellation was significant. The name stuck as successive botanists, most notably Carolus Linnaeus, adopted it in their taxonomic classifications. Plumier's act provides lots of material for pondering the nature of nature. We could quibble about which name is more accurate—talauma or magnolia. We could also wonder about the consequences of endowing what was an indigenous plant to Martinique with a French name. Furthermore, we could reflect on the impact of scientific classification on the tree's historical trajectory and its future fate. But this would all miss a more fundamental point—namely, how naming itself is an act of construction. In labeling the tree "magnolia," Plumier attached a certain meaning to the plant that was not innately part of the tree's constitution. He embedded the tree within the cognitive framework of botany as an emerging science, French colonialism as a historical experience, and to the degree that botanical classification has enabled horticulturalists to cultivate certain strands of the tree, the evolution of arboriculture in the West and elsewhere. To put it differently, Plumier did not somehow elicit the name magnolia from the plant's authentic character but rather defined that character by paying attention to certain anatomical features, and then investing them with particular meaning.

We can appreciate the social construction of trees at a more prosaic level when we notice simply how different people, in different stations in life and with various interests, see trees. In a neighborhood park, my kids and I have a favorite tree to which we often walk or bike. I don't know what "kind" of tree it "actually" is, but we are drawn to it because of its immense height. We have, in fact, named it for this attribute—we call it, profoundly, "tall tree." This may seem simple enough. But one morning we ran into a friend who also walks to the tree and who has also given it a name. She calls it "Talia," which in Hebrew, she explains, means "dew of God." Now we can pawn this difference off as mere vocabulary, yet we should also notice how the two appellations evoke distinct connotations and partially determine the kinds of encounters we each have with the tree. For me and my kids, we visit a lofty wood plant that stretches high into the sky; for my friend, she experiences the veritable water droplets of divinity. That is, she partially sees the tree as an incarnation of God. The two meanings are not incompatible, to be sure, but neither are they the same.

A similar difference explains how people relate to trees in drastically distinct ways. Many American environmentalists look into a forest and see the web of life, a precious expression of the earth's biological diversity, or fellow creatures inhabiting an organic planet. Others see timber, board feet, potential furniture or paper, or prospective land for houses, apartments, and malls. These two views are also not incompatible, but neither are they the same, and certainly they lend themselves to different forms of behavior.

As with trees, so with the birds. I mentioned earlier that if my magnolia is hard to classify as genuinely natural, it may be easier see the birds whipping around my yard as more so. Unattached to a given piece of land, they seem more easily able to escape humanity's signature. Such speculation turned out to be a ruse, as I noted how the orioles, nuthatches, and cardinals receive suet and sugar water from my neighbors, and how they are otherwise influenced by human activity. With a social constructivist sensibility, the ruse with regard to birds runs even deeper.

We know that people throughout the world have different understandings of birds. While many of us see them simply as flying creatures, others see them as models for aerodynamic study or specimens to be used in medical science. Some even understand them as divine messengers able to convey spiritual truths and principles for living.[19] Such diversity of viewpoints also differentiates our understanding of kinds of birds. To some aboriginal people, specific birds are associated with given places, formative experiences, or emotional states. The Boogoodoogada of Australia, for example, is known as the rain bird, and is not only witnessed but invoked to bring the spirit of rain into one's experience. Likewise, ever since Edgar Allan Poe wrote "Once upon a midnight dreary," many of us look on ravens with an eerie sense of foreboding. (Many residents of Baltimore look at them more warmly since the city, wishing to honor Poe, adopted the bird as the mascot for its football team.) To be sure, these are not profound differences in understanding in that they do not reveal fundamentally different worldviews. These also are not the only ways that people experience types of birds or birds in general. Nevertheless, they illustrate how much our constructions animate our understandings and experiences.

Social constructivism is relevant not simply for given creatures but also for certain landscapes, ecosystems, and nature as a whole. As Cronon points out, notions of wilderness, often a stand-in for nature, shift over time and across place. For instance, in eighteenth-century Europe and America, the word wilderness meant a place that was "barren," "savage," or "desolate." Its nearest synonym was, apparently, "waste." It represented a place in which one was prone to get lost or experience bewilderment, terror, and moral confusion. (Such connotation probably explains why Dante, a few centuries earlier, used the image of a "dark wood" as the place where he, as the main character of the *Divine Comedy*, realized his own sense of spiritual perplexity.) This notion of wilderness changed dramatically by the late nineteenth century, when many started to have positive impressions of wilderness—impressions akin to the aesthetic associated with naturalism. Wilderness, in other words, is "more a state of mind than a fact of nature."[20]

The socially constructed character of nature comes up all the time in environmental politics. One of the consistent mishaps in environmental affairs is the assumption that all parties concerned with climate change, loss of biological diversity, or ocean pollution share the same understanding of the problem. To take the most obvious example: many northern states and nongovernmental organizations work on behalf of wilderness preservation and biological diversity in the developing world, yet many in the developing world argue about the very meaning of wilderness and biological diversity at stake. In this sense, one person's wilderness is another person's home, and that which is valued as an endangered species to some is a threat, potential income, or source of food to another. The North-South divide on these issues simply underlines the more general point: nature is not a single realm with a universalized meaning but instead an ideational canvas on which people project sensibilities, cultural attributes, economic conditions, and social necessities.

To highlight the socially constructed character of nature is not, of course, to deny the reality of things other than human beings. Almost everyone recognizes that there is a fundamental substratum of materiality in the world; few of us, even the most ardent constructivists, are philosophical idealists in the sense of believing that reality is merely and exclusively a matter of mind or ideas. Drop a brick on a social constructivist's foot and it still hurts. There is genuine physicality to the world. What is less certain, however, is the *meaning* of the hurt. For this is a matter of interpretation, which itself turns on historical and cultural differentiation.

From a social constructivist perspective nature certainly exists as a materiality. Yet its meaning is up for grabs, and this is what is at the heart of constructivist ecocriticism. In this sense, we never perceive nature (or anything else) in a direct fashion, but always through discursive lenses or socially embedded narratives. As political theorist Steven Vogel notes, humans "can have no access to anything like a pre-social nature in itself; the very idea is incoherent, because all access is socially mediated."[21] Our categories for understanding nature, in other words, are social through and through, and thus nature is not separate from human life but rather part and parcel with it. This explains critic and novelist Raymond Williams's remark that "the idea of nature contains, though often unnoticed, an extraordinary amount of human history."[22]

The Social Construction of Humanity and Awakening from the Dreams

Social constructivism reveals the contingent character not simply of nature but of humanity as well. For millennia, people have assumed and argued over the essential character of human life. These efforts have tried to identify what is distinct about Homo sapiens, and understand how such unique features fundamentally animate, provide direction for, and give meaning to our lives. The search for human nature, it should be noted, is not just an empirical enterprise but also has a normative component. If humans have a nature to them, then realizing this nature should be of utmost concern to us. Otherwise, we live less than fully human lives.

Constructivism renders this enterprise problematic as it reveals the sociohistorical contingent character of human nature. Throughout the ages and across cultures, people have offered various answers to the question, What defines the human being? The list of responses includes everything from our ability to use tools, know of our own mortality, and experience self-love, to our capacity to develop language, anticipate fear, and employ reason. The problem is that none of these candidates has withstood the test of time as an absolute answer, nor does the list itself exhaust the panoply of explanations for human life across all times and cultures. This should come as no surprise. As constructivist thinkers have long pointed out, human nature, like the nature of anything, is not something inherent waiting to be discovered. Rather it is a story we tell ourselves about humanity that reflects merely particular interests, values, and passions. Put differently, Homo sapiens has no single nature or essence. Our nature, like nature's nature, is a repository for, not a producer of, meaning. We project as opposed to apperceive various natures onto our species.

To say this does not mean that human beings lack features that differentiate us from other creatures or entities. Human beings certainly do have distinct physical, emotional, and intellectual characteristics that distinguish us from the other-than-human world. Rather, it simply means that we interpret these features in particular ways, and then generalize from our interpretation to assume that these define human nature. We fail to see this as a projection to the degree that we forget the interpretative act, and assume that what we know and think about human experience is humanity's fundamental nature. Social constructivism reminds us of the interpretative act, and therewith makes clear that attempts to define human nature are not a matter of unearthing the essential character of humanity but rather merely reifying interpretations inflected by interests, values, and other aspects of cultural life.

The social constructivist orientation signals the conceptual end of nature. It suggests that we cannot naively assume that we know the true character of nature's nature or humanity's nature, since both of these are socially constructed. The conceptual end of nature, like the empirical one, then also means the demise of the dual dreams of naturalism and mastery. Looking to and counting on either nature or humanity to serve as the true, good, right, or beautiful just won't cut it anymore. Neither can serve as a fundamental informing principle for our environmental orientations or our lives more generally. We must move on.

Conclusion

Politics is not something to get rid of. The differing, often competing ends we seek do not easily compromise themselves, and life would be less rich if they did. As the dreams of naturalism and mastery fade, then, it is important to give up the desire to transcend politics, and instead ask ourselves how best to advance environmental well-being in a postnature world.

We are moving into truly dangerous times. Dramatic losses of biological diversity, the depletion of fresh water aquifers, the buildup of toxic substances in the air, water, and soil, and the increasing drumbeat of catastrophic climate change are already undermining the lives of many, and in the extreme, threaten the very infrastructure that supports all life on earth. How do we confront these in the absence of long-established categories of thought? How do we come to terms with these realities without rehearsing the battle cries of "restrict humanity's forays into nature" versus "let humanity's innovative spirit fly"? How do we live, in other words, in the absence of the dual dreams of naturalism and mastery? On what do we fasten our political understandings and engagements? Specifically, how can we respond to the ecological crisis devoid of secure philosophical and conceptual grounding?

In this chapter, I explained why the dual dreams of naturalism and mastery are no longer sufficient to guide our environmental politics. In the following pages I describe how the American environmental movement can nevertheless continue its work in the midst of such absence. I specifically show that the environmental movement need not tie itself to the dream of naturalism or fundamentally oppose that of mastery. In politics, like most else, there are middle ways that chart meaningful and promising trajectories, even if they do so without strong, secure foundations. The future of American environmentalism rests on finding such middle ground.

The movement can find a middle path by recognizing that when we awaken from a dream, many times we are still haunted or inspired by it, and this goes for the dreams of naturalism and mastery. The ends of nature may disabuse us of the theistic trappings of naturalism and mastery, but this doesn't mean that we have the ability to leave the dual dreams completely behind. Nor necessarily should we. As I will explain, when stripped of their privileged status, naturalism and mastery can indeed serve as guideposts for negotiating our way through today's complex environmental challenges. We need to see them, however, not as deterministic imperatives but merely as metaphoric poles across which to stretch our thinking. As I described in the introduction, when most of us honestly examine our own impulses, we find aspects of both naturalism and mastery alive within us. Most of us like the idea of letting nature take its course when it is convenient to do so *and* altering nature's trajectory when this becomes more appealing or necessary. Indeed, we live our lives doing both. The ends of nature invite us to welcome the practice of dithering back and forth between our naturalistic and mastering impulses without seeking permanent rest. As one might say, in restlessness is the preservation of our politics—and as I show in the following chapter, the preservation of wildness.

"Water's Freedom," Niagara Falls, New York

6

The Nature of Wilderness

In human culture is the preservation of wildness.
—Wendell Berry

Wilderness has long been dear to American environmentalism, and its preservation remains one of the oldest aims of the
movement. Since the nineteenth century, when philosophers
Emerson and Thoreau and wildlife enthusiast Muir lamented
the disappearance of wilderness under the foot of industrialization, to the movement's more modern expression in the twentieth century with the emergence of organizations such as the
Sierra Club, the Nature Conservancy, Wilderness Society, and
Wildlands Project, the American environmental movement has
been committed to protecting wildlife and wildlands from excessive human encroachment. The tradition continues today.
Each year, environmentalists in the United States (and around
the world) work to safeguard established wilderness areas in
the form of parks, refuges, bioreserves, and so forth, and seek
protection for so-called wild places that have yet to enjoy official protected status.

The empirical end of nature and social constructivist ecocriticism question these efforts as well as the long tradition
of wilderness preservation. They raise significant issues about
the meaning of wilderness itself. When most of us think
about wilderness, we have in mind a dark forest, expansive

mountain range, or deep canyon far from people, where plants and animals can live out their lives undisturbed by human beings. As I described in the last chapter, there are no such places anymore. The expansion of agriculture and human settlements, to say nothing of anthropogenic climate change, have, on the one hand, pushed wilderness to the furthest reaches of the planet, and on the other, inflected all ecosystems with a human signature such that it is impossible to now distinguish where humanity ends and nature begins, or more relevantly, where genuine wildness reigns. What should environmentalism do in such a situation? Should it abandon the effort to protect wilderness, since it is doubtful if "real" wilderness even exists any longer?

Ecocriticism adds to this challenge by suggesting that wilderness is not some universally understood condition that a body of land or water inherently possesses but rather a socially constructed idea reflecting certain sociohistorical understandings. Throughout history, few people would have defined a dark forest or deep canyon as wilderness simply because it may be devoid of human beings, and the same is the case today. Indeed, what many of us imagine as wilderness merely embodies the "received wilderness idea"—a concept that has been cultivated over the years by given societies and is meaningful to only a portion of the world's people.[1] Understanding wilderness in this way forces us to ask about environmentalists' wilderness protection efforts. What does wilderness preservation mean in light of constructivist thought? Does it matter that wilderness protection is not really about preserving a piece of land in its "given" state but instead a matter of inscribing a particular vision onto a circumscribed area?

This chapter aims to answer such questions by exploring the kind of middle path I began describing in the last chapter. Such an approach is one in which wilderness preservation is animated not by either the dream of naturalism or mastery but rather by an appreciation for how the human and nonhuman worlds have meshed into each other in a postnature age. Specifically, it explains how wilderness protection is, and always has been, a matter of human creation and management involving a tremendous amount of human thought, energy, and intervention. It also explains, however, that while not absolutely given, wilderness nevertheless has a quality of otherness to it that manifests in wildness. This chapter explores how we can hold both of these "truths" and fashion meaningful postnature wilderness policies. It looks at what principles can best inform how we can work with wilderness so that the impulse to control the nonhuman world—so central to the dream of mastery—does not get out of hand, and the urge simply to sit back and let nature do its thing—fundamental to the dream of naturalism—is also held in check. The middle way I outline, in other words, strikes a balance between the tensions of naturalism and mastery at a time when neither offers absolute insight.

The Wilderness Idea Revisited

The Grand Canyon is one of my favorite spots on earth. I first visited it when I was an undergraduate student. My friends and I hitchhiked our way to the canyon rim, and then spent four days exploring the inner canyon and the Colorado River at the bottom. That first trip provided a lasting impression of wilderness. What I remember most was what happened when we reached camp on our first day. We were staying at the Indian Gardens campsite, a little over halfway down the canyon. At one point I left my friends and walked to an overlook that had a view of the innermost canyon. The view was stunning. The river was grayish green and framed by severely inclined walls that despite the area's dry climate, were flecked with sage-colored vegetation. The air was crisp; the soundscape quiet; the view expansive.

I am unsure if it was the difficulty of the hike or the fabulous scenery, but as I sat there, something started to happen. As I perused the landscape, I found that my mind couldn't get itself around the place. My eyes had so much to look at. I could watch the river, the inner walls, or the cloudless sky. When I turned behind me to gaze back at the rim, I could see layers of vegetation zones and aeons of geologic time etched in the canyon walls and innumerable side canyons that seemingly had yet to be explored. What I especially recall was how unfamiliar and, yes, wild the whole scene seemed. The place appeared untamed and, frankly, untamable. The sheer size and otherness of the area overwhelmed not just my eyes and mind but would, I had assumed, overwhelm *anybody's* eyes and mind. No one, I thought, could get their head around something so expansive, grand, and beautiful. I remember feeling grateful to be in such a place and lucky to experience what I took to be genuine wilderness.

I have been to the canyon many times since then. I have hiked and camped inside its walls in snow, searing heat, and the cool of autumn. Over the years, I have also trekked though many mountain ranges, forests, deserts, and other so-called wild places. Every time I go, at least at one point during the trip, I have a similar experience as that of my first visit to the Grand Canyon. These places put many of us in a different space. They confront us with a world that is more expansive and radically different from that of our everyday experience. Human beings and our makings are absent from these areas, and this evokes a sense of awe as we no longer feel in control of the environment but rather sense that we are part of, and partly at the mercy of, the other-than-human world. This sense, as the romantic tradition has long pointed out, includes both a feeling of wonder and danger. In wild places the unpredictability and grandeur of the world is dramatically apparent, as is one's own sense of vulnerability to forces beyond our control. Trees can fall; creatures can attack; rivers can surge—all indifferent to human will, hope, and comfort. Wilderness, as such, produces a kind of unknowing in which many of us find ourselves inspired by the mystery of what lays beyond our ability to size things up and control the world around us. When we are in such places, to put it differently, many of us get turned on by the sense of wildness.

While many of us experience a kind of unknowing due to our inability to get our minds around wilderness areas, others obviously can understand such areas and have done so in useful ways. Indeed, the ability of some people to cognitively package these places and conceptualize them *as* wilderness areas and ones worth protecting has been critical to the received wilderness experience. That is, someone had to conceive of such areas as wild in the first place, and believe that such wildness is worth protecting. The disconnect between the received wilderness experience and the reality of people actually fabricating that experience is in fact central to the problem of wilderness in a postnature age.

Archaeological evidence suggests that until the Grand Canyon was designated for preservation in the late 1800s, people inhabited the canyon for at least ten thousand years. The earliest-known inhabitants were Paleo-Indians, who lived in the canyon's caves, and hunted bighorn sheep, deer, and even megafauna like the giant ground sloth. Later peoples included the Anasazi (known to Native Americans as the Hisatsinom), who grew beans, squash, and maize along the canyon's deltas and springs, and later still were their descendants, the Hopi, who continued to farm the green oases in the canyon until the park was officially designated in 1919. These various peoples created a network of trails from the rim to the river and across various plateaus. The Hopi and others were living on the land when Captain Garcia López de Cárdenas, the first European visitor, peered over the South Rim in the sixteenth century, and when John Wesley Powell boated down the Colorado River in 1869. And since the 1800s, generations of fur trappers, freebooters, explorers, artists, and even military commanders have lived—for a time—in the canyon. After each encounter with the canyon, people left their mark on the land and within our collective psyche—shaping our understanding, interpretation, and expectation of what we have come to call the Grand Canyon.[2]

A key turning point for the Grand Canyon and the idea of wilderness in general came in the late nineteenth century as various outdoor enthusiasts, intellectuals, and most important, politicians started to value areas of stunning beauty that were either rich in biological abundance or seemingly unmarred by human enterprise. Abraham Lincoln was critical to enshrining and packaging the idea of wilderness in the United States through his effort to protect Yosemite Valley and the Mariposa Grove of Giant Sequoias. In 1864, he ceded these lands to the state of California to be preserved as natural parts of the landscape. (In 1890, the two areas were linked and came under federal protection as Yosemite National Park.) Following his lead, successive administrations sought to lock up these so-understood unspoiled lands for recreation, scientific exploration, and public use. In 1872, the U.S. government established Yellowstone as the world's first, federally designated national park in what was then the Wyoming Territories. Other countries, like Australia, Canada, and New Zealand, soon followed suit, as they demarcated boundaries and granted protective status to a series of national parks. The Grand Canyon gained its first protection in 1893, when President Benjamin Harrison declared it a National Forest Preserve. In 1919, after going through a number of different designations—including Game Reserve and National Monument—the canyon become an official national park.

When I was sitting on the rock outcrop overlooking the Colorado River on my first visit, I wasn't thinking about any of this. In fact, I was imagining that I was alone with nature. From where I sat, I couldn't see any buildings, telephone wires, roads, or other semblances of people; neither, at that moment, could I hear or see another human being. It was as if I was gazing at the earth before or even outside of history—as if I was looking at the canyon primeval. I forgot, of course, that my experience was predicated on others drawing boundaries around the area, linking up administrative services with the state of Arizona and the federal government, and taking pains to create a certain image of wilderness. In other words, I forgot that the canyon is more than rock, trees, cactus, and river. It is also the variegated register of how various people have interpreted the canyon's walls, water, vegetation, and overall value. Over the centuries, people have seen the Grand Canyon as at once a "wasteland," "incidental landform," "vast ruin" and majestic cathedral akin to the "Sistine Chapel."[3] The canyon's meaning was never a singular understanding but rather a place that could and did hold multiple meanings—with "treasured wilderness" as only one among many. I arrived on the scene, then, not empty-headed and thus able to have a direct experience of wildness, but flecked with cultural understandings that enabled me to see the canyon through the legacy of what Harrison, Roosevelt, and the host of other nature enthusiasts thought it represented: a wild place that provides us with a so-called wilderness experience and as such is worthy of protection.

When the U.S. government designated the Grand Canyon as a national park it imagined the place had value to the degree it was largely wilderness, and that people would benefit from visiting and having the supreme wilderness experience, which meant coming into direct contact with nature rather than people. In pursuit of this image, officials actually evicted the Hopi and other native peoples who were living and farming in the area. Many of the Hopi and others were moved to what is now the Havasupai Indian Reservation, and in the process had their wickiups, gardens, and peach orchards destroyed. Wilderness meant "no people" to the crafters of Grand Canyon National Park, so they got rid of the people, and then declared the area pristine and wild. Expelling the Hopi and others is nothing new in the annals of wilderness protection. Invoking eminent domain, the U.S. government evicted native people to create the Great Smoky Mountains, Yellowstone, Yosemite, and Shenandoah national parks as well as other large protected areas such as the Adirondack Park.[4] Often, military garrisons would be sent to clear the land of human inhabitants in an effort to define and then preserve an understanding of wilderness. In one of the best books about such evictions, historian Karl Jacoby points out that squatters and indigenous people rarely left easily. Resistance was frequently long-standing, and skirmishes were common. Park officials began to see these people as enemies of nature needing to be outlawed or otherwise expelled. Inhabitants cut down trees for timber, hunted local game, and used these areas for subsistence and at times profit, and this was incompatible with the conventional understanding of wilderness.[5] Protecting wilderness, the Grand Canyon or otherwise, then, is not simply a matter of drawing and policing boundaries. It also involves marginalizing people.

The practice of getting rid of people to create wilderness is not unique to the United States. Throughout much of the world, bioreserves, national parks, and various other wilderness preservation efforts have involved evictions. For example, the Indian government removed people from various areas as far back as the late nineteenth century when it set up Reserved Forests, and it continues the practice today.[6] This includes its well-known Project Tiger, in which it not only evicts people from tiger reserves but also forbids local inhabitants from entering forests that they had long used for resources. Likewise, the San (Bushmen) of Botswana were initially included in the Central Kalahari Game Reserve, the largest protected area in Botswana, but were eventually evicted in 1997 for reasons that included the justification that "people and wildlife are not compatible in a game reserve."[7] Additionally, during the time of Banff National Park's founding, the Stoney aboriginal people were removed from the area.[8]

Evicting people to preserve wilderness assumes that there is a "state of nature," with its own rhythms and dynamics, and that pristine nature could best express itself if humans would simply get out of the way. From this perspective, human presence taints and even ruins the naturalness of a place. We embrace this view when we talk about "virgin forests" or "pristine landscapes." But if people have inhabited what are now wilderness areas for, say, ten thousand years, what does that mean for our idea of wilderness?

Humans have been altering nature in dramatic ways for centuries, if not ever since we left the Fertile Crescent tens of thousands of years ago. In both deliberate and unconscious ways we have altered the land, modified the water, and amended whole ecosystems as we made our way across the planet. We have traveled from one place to another, for instance, bringing not simply tools and supplies but also exotic species and diseases that once unleashed, devour other plants and animals as well as decimate whole human societies. Moreover, our use of fire, which we have also practiced for thousands of years, changed the landscape dramatically. The Great Plains of the United States, for example, are thought to be the result of indigenous peoples' fires set to clear the land and enable agriculture. Given this long-standing humanization of nature, it should be clear that wilderness is not merely a state devoid of people, as the U.S. Wilderness Act asserts. (The act defines wilderness as "an area where the earth and its community of life are untrammeled by man [*sic*], where man himself [*sic*] is a visitor who does not remain.")[9] Rather, wilderness is a place fabricated to look and feel a certain way. Setting up a wilderness area is in fact not just a matter of drawing a line around a given piece of land or even an area of ocean. It also involves conceptualizing, constructing, and ultimately creating a certain state of affairs—a state that has mostly entailed evicting inhabitants.

It is in this sense that we can appreciate the constructivist critique of nature and by extension wilderness. Wilderness is not some given condition that places inherently possess but instead a socially constructed one that is marked by a tremendous amount of human engineering.

Managed Wilderness

The socially constructed character of wilderness is often lost to most of us, since like my experience in the Grand Canyon attests, the "marks" of construction are usually in evidence, ironically, by their absence. Getting rid of all semblance of prior habitation makes it seem like the land has been preserved in perpetuity for centuries, if not millennia—as if forests, mountain ranges, and deserts are being seen in their most natural state. By recognizing that wilderness is an idea, we come to see that establishing reserves, refuges, and the like is not about stepping out of the way and letting nature do its thing. It represents a human choice about what lands should be preserved, what species are to be protected, and what parts of the earth are valuable in one state as opposed to another. It is about deciding where to draw lines, how to break up landscapes, and how to assign value. Wilderness is, in many ways, the antithesis of the wild; it represents something that we create, shape, and tinker with.

We see the tension between wilderness and human action in even higher relief as we realize that currently, despite it sounding like an oxymoron, we have "managed wilderness." Today's biologists, rangers, wardens, and ecologists use methods to protect wilderness areas that are themselves intensive forms of human intervention. As described in the last chapter, we have already altered the landscape so much and established ourselves as an ecological force in our own right that we can no longer simply let the nonhuman world be. Like a wild animal raised by humans, wilderness cannot survive on its own anymore, or more accurately, cannot maintain a semblance of wildness without human help. We have become, whether we like it or not, managers of the nonhuman world. We can manage this world mindfully with a concern for ecosystem health, biological diversity, social justice, and aesthetic beauty, or we can manage it unmindfully, letting the haphazard accumulation of our enterprises set our management style. Whichever route we choose, one thing is clear: leaving it alone is not an option. Today we work hard to affect a sense of wildness within wilderness areas. As mentioned, we do this at a basic level by keeping people out of protected areas as much as possible. At a more complicated level, preserving wilderness involves managing not just people but landscapes and wildlife as well. Animal reintroduction programs, efforts to eradicate invasive species, regulating water in times of drought, and feeding hungry animals in winter are basic preservationist tools these days.

In the late 1990s, biologists associated with the Colorado Division of Wildlife reintroduced the legendary lynx back into the forests of central and southern Colorado. The lynx was last seen in the area in 1973, when one was illegally trapped near Vail ski area. Reintroduction was no small task. Rather than a matter of restricting humans in the area or somehow holding back human presence, the reintroduction program required tremendous human enterprise. It involved, first, targeting and trapping lynx in northwest Canada, along the Yukon–British Columbia border. Once captured, the lynx were anesthetized, tagged, given medical exams, and transported by snowmobile to a holding facility. From there they traveled by truck, airplane, and assorted land vehicles to a second facility in Colorado, where after being fed high-protein diets, acclimated to the high altitude of the Rockies, and equipped with radio collars, they were released into two wilderness areas in the central and south-central part of the state. Wardens had difficulty getting the first groups of lynx to survive, but eventually, after much study, release, recapture, and rerelease efforts, the lynx population was showing signs of stability.[10]

The lynx program highlights the irony of wilderness preservation these days in that it demonstrates the sheer amount of human management involved with creating the wild dimension of wilderness. Not only were humans involved in the program's start-up but they also remain involved as they continue to track radio-collared lynx. It also forces us to ask about how such efforts advance wildness. Lynx are, to many, amazingly beautiful creatures. They are sleek, medium-sized cats with two distinct pointed ears that stick up as they run. Today, if you see a lynx in the woods—which would be a rare, yet in my view fortunate thing—you may notice not two, but three points coming out of their heads. In addition to their ears, you would also see an antenna connected to their collars that relays information back to game wardens, biologists, and others tracking the lynx's life. Many of us would rather see a lynx with an antenna than none at all, but the sight would be nonetheless discomforting. The silver metal wire sticking up would in some undefinable way remove a bit of the untamed character of the lynx. Extracted from their families in British Columbia or the Yukon, trapped, poked, transported, and equipped by humans, in seeing the lynx, one might get the sense that we are not looking at something fully expressive of nature but rather something partially crafted—a hybrid of sorts.

The lynx program of rewilding wilderness areas is not unique. The introductions of the gray wolf into Wyoming or moose into Michigan are further examples of wildlife management. Such management, it should be noted, involves not simply reintroducing certain species into an area but also controlling the number and distribution of species within protected areas. In 2005, for example, game wardens outside Mumbai, India, captured forty-seven leopards who had strayed from a nearby national park. They kept the leopards in captivity for over a year, and then planted electromagnetic chips into the animals' tails to allow wardens to track and recapture the leopards when they strayed too far outside park boundaries or attacked people. It is indeed common practice throughout the world to shift the population of various species to maintain the health of ecosystems and the character of wilderness. For instance, Mount McKinley Park (now Denali National Park) implemented predator control programs from 1928 to 1948

in an effort to sustain Dall sheep populations, and the Natal
Park Board moved over forty species, consisting of a hundred
thousand animals, to biologically depleted parks and reserves
throughout South Africa from 1962 to 1995.[11]

A very different though related type of wilderness management is evident in the recent efforts by marine biologists to
keep wild populations of coral alive in marine reserves around
the world. In these wilderness areas, resource managers not
only restrict the movement of boats, ships, and scuba divers in
an attempt to prevent damage to sensitive coral. They are also
building coral reefs from scratch, and in some cases, are even
stimulating their growth through electroshock therapy treatment. An important limiting factor in the growth of coral colonies is the availability of calcium carbonate. Scientists have
discovered that they can mass-produce suitable coral habitat
by dropping welded steel bars to the ocean floor and then applying a low electric charge, which draws out coral-attracting
limestone from the metal. They then graft live coral onto the
limestone-rich steel and literally grow coral at record rates.
Such coral can withstand many assaults, including warming
temperatures or changing acidity, as long as the electricity stays
on. Such efforts make one genuinely question the meaning of
wilderness. What is wild if affecting wildness involves a steady
current of electricity?

The management of wilderness is not restricted to animals.
Park, forest, and refuge managers constantly monitor plant
populations in an effort to maintain floral biological diversity
and abundance. In a unique form of plant management, naturalists dangle from three thousand foot sea cliffs on Molokai
in Hawaii to brush pollen on a flower whose only natural pol-
linator has since died out.[12] The idea of monitoring and servicing various plants and animals in an effort to support wildness
is commonplace. Keeping wilderness looking and acting wild
is hard work. Such efforts reach, arguably, their extreme when
entire ecosystems are outfitted with sensors, hooked up to

robots, cameras, and computers that together are used to oversee ecological well-being. In the "wilds" of California's San Jacinto Mountains, for example, scientists have set up an array of equipment to manage thirty acres of protected pine and hardwood forests. Devices known as motes measure light, wind, rainfall, temperature, humidity, and barometric pressure to detect the presence of a warm body or track a chill wind as it makes it way up the canyon. Completely wireless, the system enables scientists to use laptop computers to monitor continuously the "area's" ecological conditions. Similar efforts have been undertaken on the Hudson River, where in a project known as RiverNet, scientists are using dozens of instruments to track fertilizer runoff from farms, heat from power plants, and growth of algae and pollutants, and in the Florida Everglades, where water management has assumed significant technical and sophisticated human effort.

The kinds of management techniques described raise, yet again, the question of what we mean by wilderness in a postnature age. For many, the idea of managed wilderness is nonsensible, since once one intervenes in an area in an administrative or sustained fashion, the area ceases to be wild. But how else *can* we make sense of wilderness these days? Now that we humans have extended our reach to all corners of the earth and overlaid our understandings so thickly on the more-than-human world, it is difficult conceiving of wilderness as genuinely separate from our lives. Wildness, as the received wilderness idea has bequeathed it, is a seeming casualty of a more humanized world.

Wilderness in a Postnature Age

How do we undertake wilderness protection in a world in which the object itself is problematic? Here is where we see the poverty of both the dream of naturalism and mastery.

The Fading Dream of Naturalism

The dream of naturalism places nature on a pedestal and calls on humanity to align itself with the natural way of things. It assumes a clear distinction between humans and nature, and privileges the latter over the former. When it comes to wilderness preservation, we see the dream of naturalism in what is called "reservation ecology."[13] This is the notion that we can best preserve wild places by drawing a boundary around them and letting them flourish on their own. Usually it involves seeking official designation of a place to be a national park, forest, refuge, monument, world heritage site, or official wilderness area. Reservation ecology assumes that wildlands possess the ecological capabilities to approximate and maintain themselves in a "wild" state, and that leaving the earth alone ensures wildness.

A key question to ask about reservation ecology, aside from whether it makes sense in a postnature age, is whether it in fact works. There are reasons for doubt. For all its aspirations, reservation ecology has never really been able to stem the tide of humanization and inculcate sustainable practices capable of actually preserving what we take to be wilderness. Indeed, today most established wilderness areas are under siege, and the future looks even bleaker.

The pressures on wilderness are enormous. Roughly 10 percent of the earth's land is currently considered protected, although only a fraction of this can be considered wilderness as we are akin to imagining it.[14] In the United States, for example, federally designated wilderness occupies only 1.8 percent of the total landmass in the lower forty-eight states and about 4.5 percent of U.S. land if one includes Alaska.[15] You might think that given how small a sliver is reserved for the other-than-human world, people would learn to respect wilderness boundaries, to hold their gaze on this side of the divide, and let wind, rain, flora, and fauna operate independent of human will, intention, and influence. Sadly, this is far from the case. While many prize the last remaining wilderness areas, many others see them merely as places being held in abeyance, waiting for our growing numbers and rapacious appetites to demand their submission. This is the case even with legally protected areas. As we all know, governments can and often do change their minds about environmental protection and wilderness preservation specifically. This is what partially prompted Leopold to write that "wilderness is a resource that can shrink but not grow."[16] One administration may value wilderness more than another, and thus wilderness protection is frequently at the whim of changing governmental power.

One sees this with regard to ANWR. Every few years there is an attempt to allow oil and gas exploration in the refuge. Expanded in 1980 by President Jimmy Carter, ANWR represents one of the most prized wilderness areas in North America. Some call it the "Serengeti of the North America"; others refer to it as the "last wilderness"; the U.S. Department of the Interior's official Web site says that ANWR is "a place where the wild has not been taken out of the wilderness." It serves as the migration grounds for tens of thousands of caribou, and provides a home to numerous species including polar bears, whales, snow geese, and wolves. It is known that the refuge has oil deposits within it and thus there are repeated calls to open it up for so-called development. Such calls have been beaten back over the years, but there is clearly no guarantee that refuge loyalists will continue to win out. As gasoline prices increase and calls for energy independence in the United States grow louder, some predict and others hope that ANWR's days are numbered. Certainly other protected areas have had their days cut short.

Pressures on wilderness are not simply about changing governmental sentiment. There are real forces out there knocking on the door of wilderness areas, and in many cases pushing their way through. The two most significant worldwide are agricultural development and logging. Much of what we consider wilderness today is in remote areas far from the bustling, urban world. In fact, some people say it is their remoteness that has made such lands attractive for wilderness designation in the first place—their remoteness made them appear to have little economic value.[17] (In this sense, it is interesting to note that the former Bush administration's greatest contribution to wilderness protection was designating the Northwestern Hawaiian Islands National Monument, which is the largest marine protection area in the world and also the remotest.) A combination of population pressure, land degradation, increasing desertification, lack of or inappropriate land tenure polices, poor government oversight, increasing international demand for raw materials, and coercive displacement is driving farmers, loggers, cattle ranchers, and others who work the land further into the heart of wilderness areas in search of economic gain. In doing so, they often log or burn vast stretches of land with disastrous ecological consequences. By the late 1990s, according to conservation biologist John Terborgh, the combination of logging and agriculture had destroyed over half of Guatemala's Sierra de las Minas Biosphere Reserve and the Ivory Coast's Tai National Park, and over one-quarter of Mexico's Montes Azules Biosphere.[18]

The most intensive and biologically disturbing logging, burning, and land conversion is taking place in the tropics. Tropical rain forests account for only 6 percent of the earth's landmass, yet they are thought to house over one-half of the earth's species. Sadly, we are losing roughly an acre of tropical forests every second, and with this goes innumerable species. For those concerned about wildness, the decimation of tropical rain forests appears akin to a biological holocaust.[19] (And for those who live in these areas, and depend on functioning and productive ecosystems for their livelihoods, this biological holocaust challenges one's very survival.)

Many times this is happening despite governmental protection. Designated bioreserves throughout the tropics face a multitude of pressures from local, national, and international

forces. While these wilderness areas may have a legal preservationist designation, the ability to ensure that they are managed
appropriately is increasingly difficult. Challenges include, but
are not limited to, underfunding, balancing human and ecological needs, war and conflict, shifting populations, illegal poaching, and invasive species. Indeed, many protected areas lack
the staff and resources for proper security. Brazil, for example,
claims over thirty nature preserves in the Amazon. Yet only ten
of these have guards in them, and even these guards are un-
derpaid, underprovisioned, and lack the ability to effectively
patrol park boundaries. A single guard in many of these preserves is responsible for patrolling up to six thousand square
kilometers, an area larger than the state of Delaware.[20] It is no
wonder that such reserves are unable to beat back incursion
and exploitation.

All of this is not to say that wilderness preservation—in the
form of establishing national parks, forests, bioreserves, refuges, or nature preserves—is not worthwhile. That wilderness
areas are disappearing does not mean that our past efforts
have been in vain. Although many officially protected areas
have faced numerous challenges and failures in fulfilling their
purpose, it is frightening to think counterfactually, and imagine how few flecks of wilderness and the biodiversity that they
support would still be around if no protections were in place,
and if the idea of wilderness preservation itself was underde-
veloped. Nonetheless, the record of deforestation and degradation mentioned above suggests that the effort to draw a line
between humans and nature and to keep nature as pristine as
possible has severe limitations. Wild plants, animals, and lands
are disappearing at alarming rates. Traditional wilderness conservation—so-called reservation ecology—is seemingly not the
best method of retaining and protecting what we take to be
wildness.[21] Tied to the dream of naturalism, reservation ecology has ignored the consequences of human actions around
and within wilderness areas, and has placed too much faith on
wilderness areas, if left alone, to save their own skin.

Reservation ecology has a mixed record of achievement and dismal prospects for the future not only because the borders of wilderness areas are understaffed or under siege from land conversion practices. All the patrol officers in the world could not hold back many of the forces that threaten wilderness. In addition to the more direct threats of land conversion, logging, bush-meat hunting, poaching, and the like, wilderness worldwide is increasingly under threat from a host of indirect pressures. These forces do not knock at the door or somehow march to the border and push their way in. Rather, they infiltrate largely unnoticed. They come in the dark of night, as it were, and slowly but no less powerfully strip wilderness areas of their so-called wildness. Blind to park, refuge, and wilderness boundaries, air, water, soil, and species serve as mediums for transporting change across boundaries. As farms, housing developments, businesses, and industries spring up around wilderness areas, so too does the noise, pollution, erosion, and the number of invasive species. This form of threat is not so much about disrupting contiguous terrain as the subtle infiltration of humanity's multifaceted presence. It signals how our imprint on all dimensions of the biosphere influences what is going on inside the boundaries of what we continue to call wilderness.

A final and perhaps the most pervasive threat to wilderness areas is global climate change, an issue I will address at length in the next chapter. Rising temperatures are pushing ecoregions toward the poles, forcing plants and animals to migrate. This is wreaking havoc on wilderness areas as living things cannot easily migrate over roads and infrastructure—leaving wilderness areas to act like heated prisons—nor can most plants migrate quickly at all. Moreover, the melting Arctic sea ice, the infiltration of saltwater into freshwater systems as sea levels rise, and the invasion of nonindigenous species due to climate change are all exacerbating threats to wildlife and wildlands as they undermine these areas' life-support systems.

Taken as a whole, reservation ecology as a method of wilderness protection has significant limitations. It is certainly a generous and ethically upright ideal to imagine that nature, if cordoned off from human beings, will unfold along its own trajectory and therewith flourish. But in practice and theory it has much to be desired. Fortunately, as I will explain toward the end of the chapter, over the past few years there has been a growing shift from a "fences" and "borders" approach to protected areas, to one that includes human communities as well as attempts to meld preservation and human development together. In contrast to reservation ecology, this orientation has been called "reconciliation ecology" and deserves attention.

The Fading Dream of Mastery

Naturalism offers only a diminishing hope for wilderness preservation in a postnature world. What about mastery? Does the mastery narrative proffer ways of protecting wilderness after nature has ended?

At first blush, these should strike us as silly questions. The impulse to mastery has, after all, long been the enemy of wilderness preservation. The whole reason people started to care about wildlife and wild places in the first place was in response to humanity's incessant urge to control the nonhuman world. Taking charge of nature and shaping it according to human will rubbed many the wrong way, as they witnessed forests, mountains, and so forth disappearing under humanity's scientific, technological, and industrializing foot. Given this, it should seem obvious that mastery offers little to wilderness protection in a postnature age. If anything, it embraces the whole move toward leaving the wild otherness of nature behind and fabricating an even more humanized world. How can such an orientation possibly advance wilderness protection? As I see it, it can do so by emphasizing and defining the managerial dimension of wilderness protection.

A number of years ago I took my children to one of those environmental fairs aimed at inspiring people to care about and take action to protect the environment. Typically for these sorts of things, there were booths with environmental groups offering literature, demonstrations of how one could reduce one's ecological footprint, and lots of merchandise to purchase. After making our way through the exhibits and sale items we came upon a large semitrailer truck that had been outfitted inside to look like a tropical rain forest. One entered the display through the back, and walked through a short path decked with both real and fake animals as well as plants and waterfalls that were supposed to impersonate a tropical rain forest. Many of the plastic animals moved, and the sounds of chirping birds, hissing snakes, and whirling wind were piped through the exhibit. When my kids and I emerged from the truck, my then-three-year-old son asked a question that continues to perplex me: "Daddy, was that a *real* rain forest?"

If wilderness is simply a place devoid of humans and populated by plants, animals, waterfalls, and so on, can we straightforwardly dismiss fabricated rain forests—like the one on the semitrailer truck, or more challenging, the tropical rain forest building at the National Zoo in Washington, DC (equipped with "real" hummingbirds, monkeys, and red-crested cardinals)? More generally, can we dismiss human-constructed landscapes that try to affect the sensation of being in the wild? Many of us may laugh at the thought of Lion Country Safari, in West Palm Beach, Florida—with five hundred acres of "adventure," and over a thousand animals—or Arbuckle Wilderness Exotic Animal Theme Park in Davis, Oklahoma—with baboons, spider monkeys, giraffes, and other "wild" animals in habitats "similar to their own"—as wilderness areas. But it is clear that many prefer such places to more traditional wilderness preserves, and not simply for the convenience. Many people, in fact, like the wild character of them. They enjoy seeing creatures other than humans, and can do so easily in such parks or zoological gardens. Ironically, one can probably see and experience more "wildness" at such sites, as "exotic" flora and fauna—dramatically different from the kind we are used to seeing—can be kept alive and on display. By controlling plant and animal life along with the broader landscape of which they are a part, humans can cultivate something that seems like wilderness. This is right in line with the dream of mastery narrative.

It is this narrative more generally that worries little about the loss of biological diversity or deforestation. Extreme confidence in human ingenuity and technological prowess, and the faith that humanity is the be all and end all of life on earth, suggests that, if we want, we can bioengineer new species and someday even bring back extinct ones. Today, for instance, efforts are being made to preserve the fossilized DNA of extinct animals with the aim of regrowing them when we have the technical means to do so. Indeed, already researchers are finding ways to reconstitute the genomes of prehuman creatures with the hope of fabricating facsimiles of even dinosaurs.[22] In the same manner, foresters are clear-cutting forests and then replacing them with tree nurseries that will eventually grow into full-on forests. The thought is that we can simply breed animals and forests to create wilderness; we need not preserve or otherwise sustain them to do so. This goes along, of course, with a socially constructed view of wilderness and underlines that humanity is (and should be) in the driver's seat when it comes to wilderness protection. To the degree that *we* decide that we like the look or feel of places in which the nonhuman world overwhelms us, we can design and construct spaces that deliver such effect. Wilderness is not the province of the other-than-human world. It is the creation of the very-much human one.

When wilderness becomes completely a matter of design and human construction, many of us may start to wince. Yes, we may know that humans have already intervened to create the wilderness areas we now enjoy, but something still seems wrong with equating theme parks or bioengineered designer ecosystems with officially designated forests, monuments, and refuges. Trying to get at this difference can help us evaluate the contribution of the dream of mastery to wilderness protection in a postnature age.

To many of us, bioengineering a new species or completely fabricating a landscape is not a matter of wilderness protection because it gets rid of the defining characteristic of wilderness—namely, its wildness. What makes natural places, like national parks, forests, and wildlife refuges, so special is that they house otherness. Within their boundaries, the nonhuman world has a prominent presence. We may have altered the landscape, kicked people out, reintroduced species, and kept those species alive by feeding or otherwise servicing them. Nonetheless, the creatures themselves and the land or water that constitutes their habitat are still fundamentally different from us. They live lives or follow biological trajectories that while interdependent with humans, are *other* than us. They have their own way about them that is inflected by humans, but is not completely determined by us. This sliver of otherness makes them partly unpredictable and even alien. When creatures or landscapes are completely designed and created by us—through bioengineering or synthetic manufacturing—many of us draw the line and reject their wild character. Such entities seem to be no longer genuinely other but simply manifestations of human design.

There is much to this argument, but we should note where it breaks down. We may feel that bioengineered creatures or fabricated landscapes are expressions of human intention, yet we should also recognize that this doesn't make them wholly human. The mouse that "naturally" grows cancer or the cow that "naturally" produces fortified milk, for instance, still lives a nonhuman life. It acts in ways that deviate from our control. We may have created it, but we don't fully determine it. There is an element of otherness in the mouse or cow. And given this, it may not be too farfetched to understand why some people see bioengineered creatures or human-made landscapes as still wild. Such entities have an element of surprise to them insofar as living things and even the land itself can still develop or unfold along unpredictable trajectories.

To the degree that there is still an element of otherness or wildness in human-fabricated creatures, we can certainly appreciate the managerial dimensions of wilderness protection. In contrast to the dream of naturalism's take on management— where management is a concession—the dream of mastery helps us see that management could, in fact, be in the service of wildness. It could be directed toward letting the unpredictable unfold, and the complex relations between living and nonliving entities flourish. Where the dream goes awry, however, is its belief that it can completely manage nature's affairs and make nature operate according to human dictates. As I demonstrate in the next section, wildness cannot be imparted by human hegemony but must be cultivated instead. It emerges through a relationship with the nonhuman world. Such a relationship is marked neither by the arrogance of mastery nor the humility of naturalism. Rather, it involves a commitment to coexistence, coevolution, and symbiosis, and an ability to keep wonder alive.

Preserving Wildness in a Postnature Age

Postnature wilderness protection is about preserving the wildness of places in an age when our understanding and faith in both humans and nature have waned. How, then, do we go about retaining the project of wilderness preservation? Is such an aim even possible anymore? As I see it, the way forward rests not on a model of reservation or humanized ecology but rather on an ecology of relationships and responsibility. If reservation ecology is about erecting and policing boundaries, the way forward involves softening boundaries. If reservation ecology entails separation, a postnature orientation is about connection. It requires a type of wildness protection that enhances connectivity—between people, landscapes, species, and as will become clear, narratives. Having said this, it needs to be emphasized that while connectivity is important in a postnature world, the future of wilderness protection is not about completely collapsing the distinction between humans and the more-than-human world. Collapse, as such, would never protect wilderness since humanity's presence is so powerful. There needs to be some appreciation for and effort devoted to safeguarding the otherness of the nonhuman world. In other words, we need a middle way that capitalizes on the tensions between the dreams of mastery and naturalism, recognizing the many transgressions that already take place across the human/nonhuman divide. Postnature wilderness preservation entails understanding the character of those transgressions, and deliberately fashioning them in directions that augment biological diversity and abundance, reveal inherent interdependences, and most significantly, cultivate wildness.

One way to move in this direction is to blur the boundaries—both physical and ideological—that circumscribe wilderness areas. For all their constructed quality, wilderness areas still harbor wildness to the degree that the relationships and dynamics of ecosystems will always exceed the determinative and predictive power of human logic and control. Through a blurring of boundaries we can foster a greater coexistence with untamable wildness. We can effect such a blurring through the use of wildlife corridors. We can see the benefits of these in the context of land fragmentation.

In our increasingly humanized world with a growing global population and its associated demand for agricultural commodities, there is little chance that land fragmentation will slow anytime soon. Given this, while it may still make sense to establish new islands of wilderness in the form of new national parks, forests, and refuges, we have to go further. Another isolated protected area—even with better patrol forces and stricter regulation—will hold back the onslaught of the shovel and the ax only for so long. We need to find a way that employs the shovel and ax in the service of enhancing wildness. We can do this by establishing wilderness corridors, and developing ways of managing these corridors that would breathe new life into vulnerable wilderness areas.

Corridors are channels that link islands of established wilderness. They provide migration routes through which animals, and over time even plants, can enjoy safe passage away from fire, acute pollution, or invasive species, and serve as lineaments between gene pools and thus conveyor belts of genetic robustness. More generally, wildlife corridors support ecosystem health by expanding the area in which plants and animals can live. It is widely known that the more area plants and animals have to roam or live in, the greater their chances of survival. In an increasingly humanized world, we need to create safe havens for wildlife, and ways for wild plants and animals to move around or through the built environment in between such havens. Biological corridors provide one such avenue.

The idea of wildlife or biological corridors has been around at least a decade in wilderness circles. An addition to it is the notion of not only securing strips between enclaves but also softening the entire divide between networks of protected areas and their surroundings. Key to this is the idea of permeability. Reservation ecology provides a false sense of wilderness security by pretending that people can do whatever they want as long as they do so outside park or refuge boundaries. But it is increasingly clear that what happens outside these boundaries has tremendous impact on what happens inside them. This is especially the case when the borders are starkly drawn. When we build asphalt parking lots, malls, and highways, or clear-cut to the edges of established wilderness areas, we create boundaries that wildlife has difficulty negotiating. We may designate routes for wildlife to follow in the form of migration dispersal corridors, but this is no guarantee that wildlife will actually follow our directions. Oftentimes wildlife drift on to private or otherwise unprotected lands, and in these cases, it is essential that doing so does not endanger their lives. Thus, we need more than core areas that can protect wildlife from excessive human influence and linear pathways that wildlife can follow to get from one protected area to another. We also need permeable boundaries for when animals or plants roam or find themselves amid human settlement. This entails landscape design and management with an eye toward reducing obstacles for wildlife dispersal. One can imagine, for instance, selective logging rather than clear-cutting, suburban lawns landscaped to provide tree cover and habitat for wildlife, and sidewalks and driveways "paved" with materials that retain rainwater and approximate soil to encourage the encroachment of wildlife into our lives. Biological permeability, to put it differently, aims to integrate wilderness and nonwilderness areas. Such rewilding invites people to be more mindful of the other-than-human world increasingly around them, and begin to see themselves and their influence as part of nature more generally. The result is a reconsidered understanding of wildness and nature that recognizes human influence as part and parcel of so-called natural cycles.

Blurring boundaries in terms of biological corridors and landscape permeability is an important step toward preserving wildness, but it is not the only one, and certainly not the hardest. The most challenging step to blurring the boundaries of wilderness involves a cultural and philosophical blurring of our *idea* of wildness. Central to this point is the principle of relationship. For too long, humans have maintained cultural and philosophical boundaries between ourselves and wildness. From art to literature to land management, we have carried on a tradition of separation and separateness. Yet to have a relationship with someone or something, it is critical that the boundaries not necessarily come down but at least soften. It is in this sense that relationships are critical to our notions of wildness.

The idea of seeing wildness in terms of relationship finds resonance with a social constructivist sensibility. We see a piece of land (or sea) as wilderness not because of some inherent characteristic but rather because of an inherited idea and understanding we bring with us. This idea is not some static conception of what a wild place should fundamentally look or feel like but instead a matter of how humans relate to such an area. Jack Turner captures this notion when he writes, "What counts as wilderness is not determined by the absence of people, but by the relationships between people and place."[23] When we appreciate wildness as a relationship, our orientation to wilderness protection shifts. Reservation ecology calls on us to establish boundaries with the idea that by reserving a zone for nature to express itself, all we must do is hold back excessive human intervention. It assumes that we have responsibility over that which we control and essentially none over that which we don't. The problem with this approach is that it confuses the relationship between responsibility and control. While we may have limited determinative power over the natural world, this doesn't mean that we have no responsibility for it. As mentioned, we cannot simply take a hands-off approach to wilderness protection.

Many native peoples, like the Hopi of the Grand Canyon, have been consciously developing meaningful relationships between people and places for millennia. Many practice a type of caregiving in which one cultivates wildness. That is, wildness does not simply arise by our stepping out of the way or being a supplicant to natural forces but rather by assuming the role of midwife and steward. Native peoples have engaged in numerous practices that inherently reject the idea that nature must be left alone, opting instead to intervene in a way that actually shepherds it along. For example, early Native Americans actually enhanced biological diversity and ecological dynamism by selectively harvesting wild plants, intentionally burning certain terrain as well as outplanting and pruning particular vegetation. Such practices actually helped sustain not just culturally meaningful but also ecologically significant plants, such as "Indian potatoes" along with certain tobacco and cordage species. One way to think about this is that Native Americans nourished wilderness by interacting with the land in particular ways. Indeed, there is evidence that when they failed to take care of the land it would go fallow. Certain plant populations would decrease, ecological relationships would attenuate, and the variety of flora would dwindle. Such a state, while free from excessive human influence, is "different from being truly wild."[24]

One sees a similar orientation in many sacred groves in India. Sacred groves are communally protected forest fragments that have religious meaning. Communities share responsibility to protect the groves from resource exploitation even though most groves allow for limited, sustainable extraction. Evidence suggests that although populated with people, groves can harbor greater biological diversity than officially designated wildlife sanctuaries and parks. They can do so because grove communities see themselves as stewards committed to cultivating the wildness of such places. Sacred groves are believed to possess a quality of otherness, and communities undertake practices aimed at nourishing this. Groves are part of a broader set of Community Conservation Areas in India and elsewhere in which people live within forests and other ecosystems, but understand themselves to be in a certain relationship with the land such that their actions enhance biological abundance and enrichment.[25]

The idea that we have responsibility over what we cannot control should be familiar to environmentalists, especially those in the United States. In fact, one could argue that it is essential to American environmentalism's long tradition. We do not, for example, control the migration patterns of the Porcupine caribou herds of the northern slope of ANWR, nor do we dictate the journey of anadromous fish, like salmon, as they return year after year to their spawning grounds in the Pacific Northwest, Canada, and Alaska. Yet this does not relieve us of the responsibility for shepherding such migration along. Such shepherding is partially what instills a sense of interacting with

wildness. For instance, Native peoples in the Pacific Northwest clean spawning beds, open sand-blocked river mouths,
and retain sufficient water levels in an ongoing relationship
with salmon. Many environmentalists likewise work to remove
dams or install fish ladders to enable salmon to journey back to
their spawning areas. These kinds of efforts are about securing
wildness. To be sure, people undertake them in specific places,
but it is not the place itself that determines the character of effort. It is the relationship between people and the place that
does so. When we start to see wildness as a relationship rather
than simply a place, this opens up other possibilities for wilderness preservation. Wilderness areas are not necessarily devoid
of humans but instead provide places where symbiotic relationships can emerge in which people "intervene" in an effort of
mutually beneficial coexistence.

The idea of otherness underlines a final orientation toward
postnature wilderness protection: the act of opening ourselves
up to the ways in which wilderness protection can be a conversation or dialogue between humans and nature. There is no
question that wilderness, *as we have understood it*, has van-
ished. A world that is pristine, uninhabited, and unaltered by
humanity is nowhere to be found anymore. Curiously, in the
midst of such decimation, there is a slowly arising awareness
that there are nevertheless no places that are *not* wild in some
fundamental sense. As Alliance for Wild Ethics founder and
author David Abram puts it, "There is no realm, not even the
mental terrain of our thoughts, that falls completely under conscious human control." [26] This is becoming increasingly clear
to theorists of complexity and chaos theory as well as ecologists, cognitive neurologists, and meditators around the world.
Whether looking at the movements of a raindrop on a window-
pane, the human breath as it interacts with the wider atmosphere, the trajectory of forest fires, or the scattering quality of
our thoughts in even our quietist moments, things have a quality that is beyond our ability to fully map, understand, and con-

trol. What this suggests is that wildness is not simply a realm *out there* but also one *in here*. There is, to use wildlife tracker Paul Rezendes's phrase, the "wild within," which if cultivated with human creativity, craft, and attentiveness, can correlate with the wild without.[27]

One of the goals of so-called civilized life has been to develop the ability to control ourselves—our thoughts, emotions, and behavior. More generally, civilized society shares this goal insofar as its institutions prescribe "normal" or "acceptable" codes of conduct through which it can govern and organize diverse people toward common goals. We all know that despite our best efforts, however, there are always parts of ourselves that are inaccessible to conscious thought and deliberate manipulation. There are pockets of our mental, emotional, and spiritual lives that resist self-imposed or societal determinative efforts (in the same way that even the most tamed landscape has an element of otherness to it). When we experience these, we find ourselves partially off-kilter but fundamentally alive. Put differently, all of us suffer to varying degrees from what could be called a "hardening of the categories." We develop ways of seeing, feeling, and generally experiencing the world that are difficult to let go of or transcend. One of the greatest gifts wildness offers is the chance to shake off our habitual ways of being in the world and experience things anew. We enjoy this experience precisely because we are not in control.

Cultivating a sense of the wild within can enable us to care about and seek to nurture the wild without. As we come to see wildness not as a finite quality that exists in only some places but rather as an omnipresent eminence that can be tapped, accessed, or nurtured in a purposeful way, we begin to experience the pleasure and excitement of encountering otherness—in ourselves and the more-than-human world. Encountering otherness is at the most basic level what wildness is all about. Preserving places, domains, realms, and spheres of experience in which wildness' presence can be retrieved and experienced is what Thoreau, Muir, Roosevelt, Foreman, and others have long committed themselves to. We can continue this tradition in a postnature age by sensitizing ourselves to the surprises both within and outside ourselves, and finding ways of cherishing and instantiating these both symbolically and in actual material form.

Conclusion

From the environmental movement's beginnings, it has worked to keep certain areas out of reach of human influence so we could find refuge, expose ourselves to natural beauty, and experience the world outside the grip of our own hands. Since the early nineteenth century, many environmentalists have imagined wildness to be a pristine place, a region outside the controlling powers of humanity, but as I've explained, the realities of today's wilderness areas are increasingly at odds with our image and experience of wildness. This is the driving force behind the need for a postnature environmentalism, and its critical role in helping to shape new understandings and approaches to preserving wildness and wilderness protection.

Wilderness areas have been key to keeping wildness alive because they are places in which wonder arises almost spontaneously.[28] Wildness preservation, in a postnature world, needs to retain the physicality of wilderness protection as best as it can, but also needs to find ways of exercising that part of our ourselves that values and releases itself to the unbidden. Indeed, the two go hand in hand. Practically, this means that we need to supplant an ethic of reservation ecology with one of unbounded wildness. This entails continuing to create more protected areas and fortifying their security from excessive human intervention. But it also involves building wildlife corridors between existing protected areas, and softening the lines between wilderness areas and the land outside of them through landscape permeability. Furthermore, we need to cultivate qualities of relationship that are enriched by and enhance wildness. This means that wilderness management should be not only about reintroducing species, preventing erosion, and engineering the look, feel, and semblance of what we take to be wildness. It must also include, in some capacity, the experience of relatedness. It must involve the chance to delve into ourselves *and relate to the world* of which we are a part. We need to become intimate with the character of things as they unfold, even though we are part of the unfolding. While humanity's imprint is everywhere, wilderness has not been extinguished. Rather, it has simply become harder to see and cultivate.

Wilderness preservation these days, as this chapter has tried to suggest, is complicated. It involves trying to protect wilderness areas and cultivating relationships that nurture wildness. In doing this, it both appreciates the socially constructed character of wildness and paradoxically honors the received wilderness idea. Now this may sound like a contradiction to some, but such a position is unavoidable in a postnature age. At a time when we have placed the human signature everywhere and rejected simplistic understandings of nature (and by extension, wildness), we still must recognize that there is more to the world than humans, and despite the power and persuasiveness of the mastery narrative, many of these aspects escape our control. Moreover, there are some aspects of our selves that escape our control. Holding on to the wildness of life, and recognizing that doing so requires much intervention and management, is one of the paradoxes of our age.

Let me offer one final comment here. Environmental politics has long been a debate between the dual dreams of naturalism and mastery. When we turn to these visions for insight about wilderness protection, however, we see their anachronistic limitations. These limitations are wrapped up with the confidence that both dreams reveal. Both believe that wilderness can be something that we know. For naturalism, this consists of places devoid of humans—wilderness as the ultimate expression of that which we do not make. Mastery, in contrast, suggests that wilderness is that which looks most unlike us, as if wilderness is mere physiognomy. A postnature world calls for something different. Wilderness, in a world in which both human and nature's "nature" have been shattered, represents the absence of essences itself. It is its sheer otherness that escapes our categorizations and control. We engage this wildness by cultivating relationships with ourselves and others. We preserve it by developing ways for biological abundance to express itself and by keeping wonder alive.

"Wind Turbine with iPOD on Dash," Route 66, Texas

7

The Nature of Climate Change

Homo sapiens putters no more under his own vine and fig tree; he has poured into his gas tank the stored motivity of countless creatures aspiring through the ages to wiggle their way to pastures anew. Ant-like he swarms the continents.

—Aldo Leopold, *A Sand County Almanac*

Climate change is the most profound environmental challenge humanity has ever encountered. It threatens not simply to harm parts of the earth or given groups of people but to undermine the very infrastructure that supports life on earth. Already, temperatures are rising due to the buildup of carbon dioxide and other greenhouse gases, and there is evidence that this is leading to sea level rise, greater storm intensity, polar ice cap melt, and biological diversity loss.

Climate change is so profound that many associate it with the empirical and conceptual ends of nature. When McKibben declared the end of nature back in 1989, he was writing specifically about climate change. While humans had been pushing nature to the edges of the planet for centuries, McKibben asserted that climate change marks a fundamentally new moment in our relationship with the nonhuman world. It signals not just a furthering of humanity's ecological footprint but also the wholesale humanization of the earth. As he puts it, "We have changed the atmosphere and thus we are changing the weather. By changing the weather, we make every spot on earth man-made and artificial."[1] Climate change, in other words, signals that we, humans, are everywhere. There is no such thing as nature separate from humanity anymore.

Climate change also reminds us that our views of nature are themselves socially constructed. One of McKibben's key points in *The End of Nature* is that climate change not only alters the biophysical reality of the earth but the inner ideational reality of our minds as well. For many Americans, nature has long meant that which is not human. As sociobiologist E. O. Wilson puts it, nature is "all on planet Earth that has no need of us and can stand alone."[2] For McKibben, climate change fundamentally alters this understanding. By changing the climate, he writes, "we have deprived nature of its independence, and that is fatal to its meaning. Nature's independence *is* its meaning; without it there is nothing but us."[3] By interrogating nature's commonsense meaning, climate change reinforces what social constructivists have long known: that our notions of nature are not accurate perceptions of the nonhuman world but rather narratives we use to make sense of that world.

The effects of climate change on our experience and views of nature are relatively well-known. In this chapter, I aim not to rehearse them but instead to turn the issue around. If climate change has ended nature empirically and underlined the socially constructed character of nature, what does this mean for addressing climate change? That is, if there is no such thing anymore as nature devoid of human influence, and if we cannot imagine nature absent our own interpretative frameworks, how do we think about and act to alleviate climate change? Without the category of nature, on what basis do we measure the severity of climate *change* and against what do we gauge the effectiveness of our responses? Does the end of nature—empirically and conceptually—call on us to alter our mitigation or adaptation strategies? Does nature even play a role any longer in the central environmentalist issue of our age? In short, how do we think and act with regard to climate change in a postnature age?

The Boundaries of Climate Change

To repeat a quote often attributed to Mark Twain, "Everyone talks about the weather but nobody does anything about it." Little did he know that as he was expressing these words, humanity was proving him wrong. Since the Industrial Revolution, we have been pumping carbon dioxide and other greenhouse gases into the atmosphere, and removing carbon sinks through deforestation. The result is a buildup of greenhouse gases that is preventing heat from escaping the earth's inner atmosphere and changing not simply the weather but also the earth's entire climate system. Today, people everywhere are changing the climate. Frankly, not enough people are talking about it.

When we *do* talk about it, our words reveal a certain understanding. By calling climate *change* a problem, we implicitly suggest that the normal state of affairs and the good that we all want to experience is climate stability. The idea is that there is a natural, background level of greenhouse gases, and alterations away from this constitute trouble—in the form of rising sea levels, loss of biological diversity, fresh water scarcity, glacier and polar ice cap melting, and so on. This background, we imagine, is the given world. It is the world of wild nature. Nature has a way of expressing itself in climate—or put differently, there is a natural dimension to climate. Human alteration of that expression is a problem.

Climate change, thus understood, fits well with conventional environmentalism and the dream of naturalism more generally. Environmentalists have long distinguished humans from nature and have worked to minimize humanity's imprint on the nonhuman world. We humans need to reduce, stop, prevent, avoid, minimize, halt, reverse, constrain, and otherwise limit our actions.[4] We need to stay out of nature, or if we intervene, we need to do so in small, sustainable ways. In the case of climate change, environmentalists warn against humans intervening too deeply into the carbon cycle. Doing so puts too much pressure on the atmosphere's ability to absorb carbon dioxide. The atmosphere has only so much sink capacity. At some point, it fails to neutralize greenhouse gases, leading to their buildup in the troposphere.

Environmentalists' climate strategy involves trying to roll back our influence on the atmosphere. Each year we pump millions of tons of carbon dioxide into the air. At the start of the Industrial Revolution, the amount of carbon dioxide in the atmosphere stood at 280 parts per million (ppm). We have added more than 100 ppm since then, and are slated to add another 100 over the next fifty years. Carbon concentrations are growing at roughly 2.2 ppm per annum. This needs to stop. We need to slow our emissions and then ultimately halt them. We need, in short, to get out of the greenhouse gas business. The strategy for doing this seems to be policing the boundary and pushing the barricades back between humans and the atmosphere. We must keep ourselves on our side of the divide. We can do this through a variety means, including taxing carbon, setting up a cap-and-trade system, replacing fossil fuel sources, and reducing our demand for energy. Overall, the idea is to cut back and ideally halt our incursions into the atmosphere, with the aim of avoiding the most catastrophic effects of climate change and, over the long run, giving the atmosphere a chance to cleanse itself of anthropogenic greenhouse gases. In that distant future, we will, once again, establish climate stability. This doesn't mean that the earth will immediately build back the polar ice caps, lower ocean water, or restock its biological diversity. But it is believed that eventually these too may right themselves. One day we might get back to an earth that controls its climate rather than one in which humans are at the steering wheel.

Climate stability is certainly an appealing aim. Stability allows predictability. We can grow food, build secure homes, plan events, and on the whole, live freer lives to the degree that we are not beholden to a constantly changing climate. Moreover, it has the appeal of representing an atmosphere devoid of human signature, an atmosphere without additives. A key question encouraged by the end of nature arguments, however, is whether such a situation is really possible. Can we get out of the atmosphere-altering business, and if we do so, can we expect the atmosphere to deliver climate stability? There are serious reasons for doubt.

First, as we all know but often choose to forget, climate stability is a relative phenomenon; it makes sense only within certain time frames, and even within these circumscriptions, it is difficult to identify. Viewed in geologic time, climate stability is basically a myth—something that climate skeptics are all too happy to point out. Average global temperatures have oscillated over millennia due to wind patterns, oceanic circulation, sunspots, and even periodic fluctuations in the earth's orbit. One of the largest warming episodes occurred roughly fifty-five million years ago in a period known as the Paleocene-Ecocene Thermal Maximum (PETM). During the PETM, the earth's temperature rose roughly nine degrees Fahrenheit (five degrees Celsius) in less than ten thousand years. This led to mass extinctions in the oceans and shifts in plant communities due to altered rainfall. The PETM so dramatically altered the earth's ecosystems that it is believed to have helped usher in the "Age of Mammals."[5]

There have been, of course, other global warming episodes. Paleoclimatologists tell us that these episodes happen roughly every one hundred thousand years—interglacial periods in which the globe warms for about twenty thousand years and then regresses into a longer ice age. In fact, we know that we are now in the latter stages of an interglacial age that began about fifteen thousand years ago. The grand sweep of this global warming has nothing to do with human activity. Rather, it is a "natural" event in the planet's oscillating dance between ice ages and warming periods. Climate in fact has never been stable. There was never a time in which the earth was not heating up or cooling down. This raises questions about environmentalists' climate orientation. If there is no baseline for climate—if there is no natural temperature level inherent to the earth itself—then how can environmentalists justify their strategic aim? Doesn't the whole goal of aiming toward climate stability appear chimerical?

Environmentalists are certainly aware that climate stability is a myth when measured over the millennia; they rightly insist, however, that it is far from mythical within any meaningful, human time frame. Climate has been relatively stable throughout Homo sapiens' tenure on earth. Average global temperatures have fluctuated by less than one degree for the last ten thousand years. Climate stability for environmentalists does not mean some absolute state but simply a livable one, and they identify this with the past two thousand years.[6]

It is the same thing with carbon levels. From about 650,000 years ago until 1750, with the start of the Industrial Revolution, carbon concentrations fluctuated but did so within a fairly narrow range. According to the Intergovernmental Panel on Climate Change (IPCC), the authoritative body on climate science, carbon levels stood at roughly 260–290 ppm for most of humanity's time on earth.[7] The worry is that today, they stand well above this range. Current concentrations hover around 390 ppm, and as mentioned, are rising quickly. Given the association between present-day carbon levels and contemporary climate disruption and the projections of this association in the future, most environmentalists are not seeking some objective level of climate stability but merely something approaching that of preindustrial times. *This* would be meaningful climate stability.

The problem is that even this moment of climate stability is unreachable. The IPCC makes it clear that we have already passed the threshold at which we might have been able to reverse course and recapture a preindustrial atmosphere. Even the most ambitious proposals for reducing carbon dioxide acknowledge that 270 ppm is a fantasy. Many of the most concerned identify a limit of 450 ppm of carbon dioxide as the key tipping point and call on humanity to stay on this side of such a line. They warn that once we cross that threshold, climate catastrophe is all but certain. (Some, like those associated with the Step It Up Campaign and 350.org, say that we need to get below 350 ppm to avoid catastrophe.) Given this, nature or carbon stability—represented in the form of preindustrial carbon levels—provides little guide for addressing climate change since it stands as an impossible goal.

Nature recedes into the background even further as we realize that returning to preindustrial carbon levels has been out of reach for decades. When NASA scientist James Hansen first testified before the U.S. Congress in 1988 and sounded the alarm over climate change, we were already too far beyond the possibility of avoiding climate change.[8] There is a time lag between the release of carbon dioxide and its accumulation in the atmosphere. Had the world in 1988 immediately halted its use of fossil fuel and stopped deforestation—essentially cutting carbon emissions to zero—the prospect of returning to a so-called natural atmosphere would still be fantasy.

The point is that we are now in new terrain. By releasing billions of tons of carbon dioxide since the beginning of the Industrial Revolution, we have left all semblance of an atmosphere unadulterated by humanity far behind. Getting back to nature is not an option. Climate change thus can't possibly be addressed through the lens of holding an ideal of naturalness in mind and remaining steadfast toward its realization. It is, in this sense, immune from the dream of naturalism. We have already remade nature when it comes to climate, and we would do best to realize this. Climate politics must now be about fashioning the kind of nature we think best suits us and the other creatures on earth given contemporary climate realities. To engage in this kind of enterprise, environmentalists need a new narrative. We must recognize that nature cannot serve as our north star toward which we tack in times of environmental challenge. Clouds have overtaken the night sky of environmentalism. We are on our own to define and negotiate our way toward a collective destination.

We may be unable to chart our course according to nature, but surely we can navigate our way around genuine dangers. We can do this, many tell us, by avoiding the 450 ppm threshold. Up until 450 ppm, while Greenland's ice sheet will partly melt, the North Atlantic Gulf Stream may still slow, sea levels will still rise, storms will still be more intense, and biological diversity will still plummet, vast numbers of us may still be able to get through our days with some semblance of contemporary life. Over that limit we bake, however. At 450 ppm, we throw the entire climate system and many of the earth's ecosystem functions out of whack. We get runaway climate change—change happening so fast that the very biochemical infrastructure of many organisms will be pressed to the limit with difficult prospects of adaptation. Moreover, the various ecosystem changes associated with erratic climate—sea rise, extreme weather, and so forth—will feedback on each other even more and set in motion harmful changes for centuries, if not millennia, to come.

Fixating our gaze on 450 ppm makes lots of sense. The scenario just described doesn't sound attractive. In fact, it appears downright nightmarish. We should do everything we can to avoid it. But that doesn't clear up the issue of limits and the stability we seek. What does the figure of 450 ppm really represent? We call it a threshold, but for whom and in what ways is it? Here is where the socially constructed quality of climate change, and by extension nature, raises its head in a profound way.

The 450 number represents some type of limit, and environmentalism knows a lot about these. In this case, the limit has to do with the earth as a carbon sink. Excessive amounts of carbon dioxide are overwhelming the absorptive capacity of the land and oceans, and hence accumulating in the atmosphere. But where exactly is the limit in this? We are already witnessing hotter weather, the melting of everything frozen on earth, the loss of biological diversity, and so forth. Does this mean we've reached a limit? It all depends on *who* we are and how much ecological change we are willing to accept.

Certainly the victims of Hurricane Katrina, those caught in the European heat waves of the past few summers, homeowners in the path of the California, Siberia, and Greek fires, the Alaskans of Shishmaref who evacuated their village before it was lost to the Chukchi Sea, and similar villagers of Malagisa in Papua New Guinea who lost most of their land to the Solomon Sea would all say that we've hit a limit. Likewise, we have reached a genuine limit for the numerous plants and animals that have disappeared because of climate change. The 450 ppm figure probably means little to these people and the other-than-human animals and plants that have already suffered or died due to climate change.

And yet we know that the kinds of weather disruptions and suffering involved so far are child's play compared to what is coming down the pike. Millions and probably billions of people will be at physical risk as climate change continues. At the extreme, whole swaths of humanity will be wiped out as the oceans rise to nightmarish heights, food supplies dwindle due to longer and more intense droughts, and fresh water becomes mind-bogglingly scarce. But does even this scenario constitute a limit? It all depends on how much of the nightmare we are willing to live with. For many of the rich, there is the belief that enough wealth will enable one to ride out the adversities of climate change. The view that greater economic capacity can enable people to adapt comfortably to climate change is in fact relatively widespread.[9] Already people are envisioning survivalist technologies aimed not at mitigating climate change or collectively adjusting to it but rather at allowing individuals to sequester themselves from its most ferocious consequences. Judging from much of the world's inaction on climate change, it seems that too many of us are willing to live with frightening prospects.

Reservation Climate and the Fading Dream of Naturalism

The effort to draw and police a line between humans and the atmosphere with regard to climate change should sound familiar. Environmentalists have long tried to protect nature from humans. We see the same thing in relation to climate that we do with regard to reservation ecology—the practice of cordoning off wilderness areas and safeguarding them from human influence. Climate protection, sensibly, seems to consist of keeping human intervention at bay: drawing a boundary around the atmosphere, and calling on humans to stay out. We see this strategy in most of environmentalism's efforts to cap, reduce, shrink, or diminish humanity's carbon footprint. How well has this strategy worked? Has it minimized climate change? Has it protected us from the dangers of sea level rise, loss of biological diversity, more intense storms, and so forth? Does circumscribing the atmosphere and calling it off-limits to humans work?

Frankly, it is hard to say. We have never really given the idea a genuine shot. Environmentalists have long called for getting out of the greenhouse gas business, but the world has never found the political will to actually implement such a strategy. This failure itself, though, calls on us to question the orientation as a whole. At the international level, the Kyoto Protocol represents the world's guiding strategy for reducing carbon emissions. The protocol, which emerged out of the 1992 United Nations Framework Convention on Climate Change, was signed in 1997 and came into force in 2005, when over 55 percent of the main greenhouse gas–emitting industrialized countries ratified the treaty. Kyoto commits individual nations to aim for specific carbon reduction targets, with a 5.2 percent average reduction for the thirty-seven developed countries party to the protocol.

Kyoto is an important document. Representing decades of scientific research and political negotiation, it stands as the most codified of approaches and commitments to climate protection.

A key problem with Kyoto, however, is that while aiming at 5.2 percent is admirable, such a goal is far from what we need to avoid the most catastrophic, large-scale climate disruptions that scientists predict. In its 2007 report, the IPCC warned that industrialized countries would have to cut emissions 80 to 95 percent by 2050 to limit carbon dioxide concentrations to 450 ppm. The world as a whole needs to reduce emissions by 50 to 80 percent. In this light, while Kyoto may represent political reality, in terms of what has been possible in an international negotiation context, the protocol is out of synch with biophysical realities. Even if the United States had joined the Kyoto process and all the signatories met their commitments, the effect would be minimal. Some argue it would lower temperatures in 2050 only by about 0.1 degrees Fahrenheit. This represents a postponement for less than three years of temperature increases.[10]

Of course, Kyoto was never expected to be an answer to the climate challenge. Most see it simply as a step along the way toward a more ambitious international treaty. Indeed, it is scheduled to expire in 2012, and countries are now negotiating a post–Kyoto Protocol. This is good news, since the majority of the signatories have not even come close to fulfilling their Kyoto commitments. Greenhouse gas emissions have increased in almost every case, and the only significant dips have been due to economic downturns. Given our experience with Kyoto, it would be good to reconsider the overall framework of our approach as we move beyond the protocol and realize that we are in a postnature world.

Much of the thinking behind Kyoto involves restricting carbon emissions by asking people to cut back on their energy use. While people can do so through technological advances that call for little behavioral change, many see significant behavioral change as indispensable to any genuine effort toward carbon reduction. People simply must be willing to give up some of the pleasures associated with energy consumption. They must be willing to sacrifice their energy consumptive desires. Such an orientation is in keeping with environmentalist understandings.

Environmentalism has long peddled the idea of sacrifice. It has traditionally called for curtailing our desires, constraining our material impulses, and limiting our presence on the earth in the service of global environmental protection. Conservation, preservation, and sustainability, the guiding lights of environmentalism, all counsel such forfeiture. They recognize that humanity cannot continue to increase its numbers and cultivate insatiable appetites for material goods on a finite planet without consequences. The dream of naturalism has informed this orientation in that aligning ourselves with nature requires self-constraint. To respect, follow, or even enjoy nature, there needs to be a more-than-human world out there. For that to happen, humans must make some room for other creatures to live out their lives, and for the earth's ecosystems to benefit from biological abundance and ecosystem functionality. The question we need to ask is how defensible such an understanding is after nature and in the face of accelerating, dramatic climate change. With the category of nature unable to play a powerful conceptual role—toward which one aims—does sacrifice, as a political category, still make sense, and does it do so specifically with regard to climate change?

Reducing our numbers and the amount of material throughput in our lives will certainly decrease the quantity of carbon we pump into the atmosphere. Yet it will not stop it, and this is critical to understand. Conservation, preservation, and sustainability are all premised on the notion of cutting back. They counsel us to stay out of nature, and when we must engage it, do so in an almost miserly way. We should take only what we need and over time reduce our needs. The problem is that reducing our consumption will not solve environmental problems and cannot in itself save us from climate change. It can slow down the onslaught of climate dangers, and this should not be underestimated as a transition strategy. But it does nothing to create a system that will one day avoid them altogether.

Architect William McDonough and chemist Michael Braungart, authors of the book *Cradle to Cradle*, make this clear in their critique of conventional environmentalism. They criticize environmentalists for being too wedded to the idea of curtailing consumption, and preaching a mantra of reduction, constraint, restriction, and diminution. Such an orientation can never be an answer to our ecological woes because it calls on us simply to be what McDonough and Braungart call "less bad" rather than "good." When a company cuts its emissions of cancer-causing chemicals by 70 percent, for instance, this doesn't protect us from cancer; it just postpones the amount of time before more of us will be poisoned. Similarly, when we decrease our carbon footprint by using less fossil fuels, this doesn't stop the buildup of greenhouse gases but instead merely delays climate dangers. As mentioned, even if all countries met their Kyoto commitments, this would postpone increases in temperatures by a mere three years. As McDonough and Braungart put it, reduction itself "does not halt depletion and destruction—it only slows them down, allowing them to take place in smaller increments over a longer period of time."[11] In terms of climate change, cutting back our individual and collective carbon footprints will still get us to the "end of the world"; it will do so, however, at a slower pace.

Sacrifice as an environmentalist strategy for climate change is also morally problematic and out of synch with the trends we are seeing around the world. After centuries of experiencing abject poverty, many Chinese are beginning to enjoy the fruits of material abundance. For the past decade, China's economy has grown at roughly 10 percent a year. This has enabled its people to consume more resources and produce significantly more waste. China now uses more steel, grain, and water than any other country. It is the top producer of pigs, importer of oil, and emitter of carbon dioxide. Conservation, preservation, and the like call on the Chinese to curb their appetites, to want and use less. But how fair and realistic is this? After experiencing economic development premised on using nature's resources and generating waste without abandon, is it fair for northern environmentalists to advocate smaller material diets for China and other countries of the developing world? Furthermore, it is even pragmatic to do so?

A couple decades ago, the average Chinese person used the equivalent of a sixty-watt lightbulb's worth of energy for a half hour per day. This relatively minimal amount kept much of China literally in the dark, and restricted many of its citizen's access to basic needs, and certainly many comforts and pleasures. Today, China is building roughly one new coal-burning plant every week to fuel its accelerating economy and adding fourteen thouand new cars to its roads each day. Its citizens are buying iPods, refrigerators, plasma televisions, and other energy-consuming products. While China has every right to boost its country's energy consumption, especially since its per capita consumption has yet to equal that of more developed countries, the planet cannot stomach a fossil-fuel-gorging China. It is unsustainable from a global perspective. But can we really preach a message of resource reduction to the Chinese when they are finally and simply coming on line as consumers? What right do we have to do so?

Independent of moral qualms, moreover, we need to ask about realistic prospects for such a message to be heard. China is increasingly becoming aware of the damage it is doing to its air, water, soil, and species. As a result, it has established relatively high mileage per gallon standards for its vehicles and outlawed plastic supermarket bags to reduce its use of petroleum as well as committed itself to building more efficient manufacturing and energy-producing plants. Still, how much will these changes matter given the number of Chinese and their growing consumptive habits? China's cars may get better gas mileage than those in the United States. But as more families purchase vehicles, the ecological benefits of those standards get washed out. If car ownership in China ever reaches that of the United

States, there will be 1.1 billion more vehicles on the road. This will produce more carbon dioxide annually than the rest of the world's transportation systems and consume more oil than the world presently produces.[12] What can reduction really mean in the Chinese context?

The most devastating critique of an environmentalist message of sacrifice and the attempt to get out of the greenhouse gas business through reduction may be its overall sensibility. Conservation, preservation, and sustainability identify the problem of climate change as people. We are the culprits—specifically that part of us that wants to grow, consume, and materially experience the many facets of life. This is that part of us that thrives on human ingenuity and the breaking of various kinds of barriers. Cutting back calls on us to curtail this dimension of ourselves. It asks us to turn off the urge for more material things, greater convenience, and more ease. In this sense, the message of sacrifice has an almost misanthropic ring to it. It suggests a dislike of humans, since our numbers are so high and our material acquisitiveness (and its accompanying energy use) is so significant. Misanthropy is a difficult politics to advertise and sustain. It offers, at best, a doom-and-gloom mentality in which we see ourselves as fundamentally the source of our problems.

The message of sacrifice has this character partially because it comes out of the broader dream of naturalism. The dream sees humans as essentially subject to nature and understands our environmental task to be one of shrinking our ecological footprint to fit into nature's constraints. As outlined in chapter 3, environmentalists propose this out of a combination of prudential, principled, and aesthetic concerns. Such a view, while noble, is difficult to sustain anymore. Now that humanity has extended its presence everywhere, the idea of situating ourselves within nature's constraints is hard to understand. Furthermore, now that our notions of nature are themselves contestable, it becomes harder to hold nature up as an ideal toward which our lives should be targeted. In climate change, like most other environmental challenges, we need a new orientation, a new politics.

Geoengineering and the Poverty of the Mastery Narrative

There are many who appreciate the problems with cutting back and have proposed alternative orientations. Many of these draw on the dream of mastery. One of the most dramatic is that which insists that since curtailing our practices is so morally and pragmatically challenging, we should extend ourselves even deeper into the nonhuman world and fabricate our environment. This involves not "fitting into" nature's constraints but rather realizing that we are the governors of the other-than-human world, and that this gives us license and responsibility to shape it as we see fit. This orientation resonates with the dream of mastery subscribed to by critics of environmentalism and environmental skeptics. In the case of climate change, this sensibility has become attractive not simply to antienvironmentalists but also many environmentally minded people who have become simply too scared about the prospects of climate change absent significant human action.

Reservation ecology defines climate change as the buildup of too much carbon dioxide and other greenhouse gases in the atmosphere. It takes the atmosphere's ability to absorb such gases as a given and recommends reducing the amount we emit. One could imagine, and many are now doing so, defining the problem in a different way. One could argue that the problem isn't that we are pumping too much greenhouse gas into the atmosphere but rather that the atmosphere is not absorbing all that we are emitting. That is, the problem is not that we are doing too much—pumping too much carbon into the atmosphere—but that we're doing too little—we are not working hard enough to alter the atmosphere. We are failing to retrofit the atmosphere to accommodate our lives; we are failing, in other words, in our effort to master nature. Responding to climate change, according to this line of thinking, involves not shifting our ways of life but rather maintaining them by extending ourselves deeper and further into the nonhuman world, and enabling its oceans, glaciers, and the like to withstand the assault we seem determined to sustain on the earth.

Today, various thinkers are dreaming up ways to extend ourselves deeper into the geophysical dynamics of climate change. Much of this comes in the form of geoengineering. Geoengineering involves deliberately altering the earth's ecosystemic infrastructure. Geoengineers are proposing a host of schemes. One approach employs ocean iron fertilization to pull carbon dioxide out of the atmosphere—a practice that would involve further colonizing the oceans to perform greater tasks for us. Specifically, it aims to sequester carbon dioxide in ocean sediment and bottom currents by stimulating phytoplankton growth. Phytoplankton, like all plants, absorbs carbon dioxide. The more that can grow, the more carbon dioxide the oceans can sequester. Iron stimulates phytoplankton growth. Thus, the more we put into the oceans, the more carbon dioxide–eating organisms we will have. One of ocean iron fertilization's most outspoken advocates, oceanographer John Martin, once remarked, "Give me a half tanker of iron, and I will give you an ice age," referring to the idea that ocean iron fertilization could reduce carbon dioxide enough to lower temperatures and cause a mini–ice age.[13] Present-day geoengineers have their eye on high-nutrient, low-chlorophyll regions such as the Southern Ocean and various parts of the Pacific Ocean with the hope of sequestering several hundred million tons of carbon dioxide annually. A California start-up company called Climos plans to test iron fertilization in a four thousand square mile area of the ocean.[14]

The oceans are not the only domain being targeted for increased human intervention. The atmosphere itself is apparently ripe for greater human presence. One idea is albedo enhancement. A combination of cloud cover, atmospheric gases, and aerosols partially reflect sunlight from hitting and warming up the earth's surface. It is estimated that this albedo effect sends roughly 30 percent of sunlight back into space. Chemist and Nobel Prize winner Paul Crutzen, environmental scientist David Keith, and others have proposed adding to the earth's reflectivity by injecting sulfur dioxide aerosols into the atmosphere. The idea for this emerged when Mount Pinatubo erupted in 1991. The volcano spewed millions of tons of sulfur dioxide into the atmosphere along with billions of tons of magna. For months, the surfur dioxide blocked out sunlight, leading to a 0.5 degree Centigrade (0.9 Fahrenheit) decrease in global temperatures. Anthropogenic sulfur dioxide forcing seeks to mimic this and thereby induce a type of global dimming. Crutzen proposes depositing sulfur and hydrogen sulfide (which react to form sulfur dioxide in the atmosphere) into the air using artillery shells and balloons. Keith and others are less specific about how to get such substances into the air, but share the sense of promise. To be sure, albedo enhancement advocates do not encourage such action as a matter of course. Rather, they offer it as a stopgap measure in the midst of rapid climate warming. They claim that there may come a time when our efforts to mitigate and adapt to climate change have been exhausted, and we need quick relief from the most devastating effects of climate change. When this scenario comes about, they argue, few will resist albedo enhancement or other geoengineering ideas.[15]

A variation on the theme of blocking out solar radiation is space sunshades. Instead of using sulfur dioxide to screen out sunlight, some researchers have advocated an array of space optics that would deflect about 2 percent of the sun's rays away from the earth. Plans propose launching optics—which would consist of trillions of small, free-flying spacecraft—one million miles above the earth into an orbit aligned with the sun, called the L-1 orbit. Together these spacecraft would form a long, cylindrical cloud with a diameter about half that of the earth and about ten times longer, which would somehow stay between the earth and the sun.[16] Astronomer Roger Angel estimates that spaceshades would cost a few trillion dollars or 0.5 percent of world's gross domestic product—a fraction of the cost of climate catastrophe.[17]

A similar proposal involves increasing cloud cover over the oceans. This could be done, so we are told, by atomizing seawater to create greater evaporation, to in turn create clouds and intensify the density of existing clouds, therewith increasing the earth's albedo effect. In one version, wind-powered, satellite-controlled ocean vessels would disperse particles for cloud formation across certain parts of the ocean. Some estimates claim that an increase of 4 percent in cloud cover over the oceans could significantly mitigate global warming.[18]

Not everyone is keen on reformatting the infrastructure of the planet along the lines of geoengineering. Others are happy enough with reengineering many present-day creatures and plants to adapt to climate change. For instance, scientists are designing plants that will absorb more carbon dioxide, resist drought, and flourish in changed and changing ecotones. Others are working to genetically manipulate cows and sheep so that they will not emit methane as well as to alter their diets to decrease belching. Still others are proposing artificial "trees" that would strip carbon dioxide from the air. These trees would pass air through filters containing sodium hydroxide, which reacts with carbon dioxide to form sodium carbonate. The sodium carbonate could then be piped away into vast storage reservoirs under the seabed in an act akin to schemes of carbon sequestration.[19] (Of course, carbon sequestration itself is a form of geoengineering insofar as it entails digging huge holes in the ground and filling them with excess carbon dioxide.)

Behind all of these proposals and designs is the idea that contrary to a conservation-minded approach to climate change, we need not restrict ourselves from delving into the nonhuman world but rather should vamp up our ability and desire to do so. Geoengineers see us as able to have it all: grow economically, use fossil fuels, increase our energy use, and the like. We need only set our minds to manipulating the world around us that much more. We need only to advance, even more powerfully, the modernist dream of controlling and manipulating the nonhuman world.

Protecting Climate in a Postnature Age

Geoengineering may hold some promise, but it is also pretty distasteful in that it reeks of a kind of arrogance that makes many of us uncomfortable. In fact, it expresses a sensibility that many feel got us into our climate woes in the first place. And yet it should sound fairly familiar. We have been practicing it for centuries. Geoengineering is not some new technique that we have cooked up as a response to climate change. It *is* climate change. By pumping billions of tons of carbon dioxide and other greenhouse gases into the atmosphere, we have been manipulating the earth's fundamental infrastructure. To be sure, we have been largely unconscious of such fashioning, but this doesn't dismiss the result. By extracting and burning fossilized life from under the earth's surface and denuding the planet of its forests, we have altered and continue to alter not just the atmosphere but also the oceans, land, and other species. We are acidifying the oceans, forcing ecotones to migrate by shifting temperature zones, and wiping out inordinate numbers of species as anthropogenic climate change increases. It is not as if a handful of geoengineers are coming out of the closet in the face of climate change. All of us, for centuries, have been geoengineers. We have simply neglected to put this on our resumes.

In this sense, geoengineering offers a point of departure for thinking about postnature climate protection. It is untied to any particular—especially idealized—notion of nature and recognizes that any future world will be one largely of our making. Geoengineering's blind spot, however—and this is why it appears distasteful to many of us and why it can only serve as a point of departure—is its false belief that such a future will be wholly of our making. It too tightly embraces the dream of mastery and counsels a rather old slogan of how to get out of our collective dilemmas: "If brute force doesn't work, you're not using enough of it."[20] The problem for many of us is that brute force and the urge toward mastery that it represents is at the root of climate change, not an answer to it. That is, mastery brought us the nightmare of climate change. We should not rely on it solely to save us.

Where do we go from here? If neither the dream of naturalism or mastery is clearly useful for addressing climate change, how do we develop a postnature climate policy or philosophical orientation? There are a number of ways of thinking about climate change protection that make sense in a postnature age that borrow some of the impulses and insights of geoengineering without embracing geoengineering's hubristic sensibility. Each of these has been emerging within the environmental movement, but my hope is that by bringing them into high relief and contextualizing them within a conversation about the empirical and conceptual ends of nature, they will gain greater attention. These represent steps that can move us away from a restoration ecology orientation toward climate protection without embracing the geoengineering ideal. If traditional climate politics has been about erecting and policing boundaries between humans and the atmosphere, these steps are about softening such boundaries. If restricting human intervention into the atmosphere is about separation, these steps involve connection. They advocate a type of climate politics premised on connectivity—between the human and nonhuman worlds. Like the boundaries constructed and maintained by well-meaning wilderness advocates, the boundaries that have come to define climate change and climate politics are being transcended all the time. A postnature climate change politics seeks to understand the character of these transgressions, and deliberately fashion them in directions that enhance our sense of care for one another and the nonhuman world of which we are a part. Like connectivity in wilderness protection, climate connectivity, while about softening the boundary between humans and nature (and in this case, humans and climate), does not entail erasing the boundary completely. The geoengineers of the world neglect the otherness of the more-than-human world and assume that we can simply imprint ourselves everywhere with few negative consequences. A postnature climate strategy must not go fully in this direction. Rather, we need a middle path.

A first step involves moving beyond a focus on thresholds and boundaries, and concentrating on the kinds of lives we would like to cultivate together. These days, climate change politics consists of scientists proposing particular limits—maximum ceilings on greenhouse gas emissions, acceptable temperatures, or livable carbon dioxide concentrations—and environmentalists using these numbers to advance certain policies. The idea is that we need to respect the atmosphere's biophysical limitations *or else*. I firmly believe this, but also recognize its political drawbacks. For example, such strategies in general have a mixed historical record, and this in itself should give us pause in embracing them. The modern environmental movement of the 1960s and 1970s was full of dark prognostics in which various authors and activists warned that unless we could rein in the human species, we would face almost certain ecological apocalypse. The environmental movement lost credibility in embracing such dire warnings in the form of thresholds across which we were warned not to go. The failure of many scenarios to materialize led some observers to equate environmentalist warnings with old-time Malthusianism, which focuses solely on supposedly *given* biophysical thresholds without attending to human resourcefulness and our ability to bend ecosystem constraints. As I see it, it is precisely our resourcefulness and ability to bend conditions—including and especially social ones—that hold promise for confronting climate change.

Climate change is not simply a calamity but also an opportunity, and making this clear is at the heart of a postnature climate strategy. When we advertise thresholds and limits, we often do so through a language of fear. The message is: if we mess too much with nature—in this case, the carbon cycle—we invite ecosystem catastrophe. Behind this message is the old politics of sacrifice and the dream of naturalism. A different message is given and a different style of political motivation is employed when we cease projecting thresholds, and instead start envisioning and building together economic, political, and cultural systems that are simply more attractive than the status quo—in terms of economic promise, democratic governance, resilient and meaningful communities, and international security. In this sense, climate change may just be the kind of challenge many of us have been waiting for.

Recently a spate of research is demonstrating that climate change options, far from being expensive, socially dislocating, or more dangerous than contemporary practices, actually offer many benefits. Transitioning to a clean energy economy promises to jump-start or revitalize weak economies, and set even sound ones on firmer ground. Shifting to renewable energy sources will create green-collar jobs, reinvigorate former manufacturing sectors, drive significant capital investment, stabilize energy prices, and open up whole new professional enterprises associated with carbon accounting, policy analysis, and energy engineering. According to a joint study by Worldwatch Institute and the Center for American Progress, renewable energy creates more jobs per unit of energy produced and per dollar spent than fossil fuel technologies do, and such job creation will have a great impact on both the U.S. and world economies.[21] We already see some of this promise in the growth of various renewable energy industries. Global wind energy generation, for instance, has more than tripled since 2000; the production of photovoltaics has more than sextupled since 2000; and although controversial, the production of fuel ethanol from crops has more than doubled since 2000 and is slated to grow dramatically over the next few years. Add to this the amount of investment being channeled into renewable energy more generally, and it appears that addressing climate change is not some dour prospect marked by certain economic decline but rather an opportunity to fashion a more productive and ecologically sound economy. As Congressperson Jay Inslee and political analyst Bracken Hendricks put it, "This is not about sacrifice; it is about economic growth, productivity, and investment."[22]

A similar point has been made with regard to climate change strategies and national security as they relate to oil supplies. Much of the world's oil comes from a handful of producers, and many of these are military flash points. Of the ten countries with the largest proven oil reserves, only one is genuinely democratic—namely, Canada—and most of the others are security tinderboxes. Today, conflicts abound in Iraq, Columbia, Angola, Nigeria, and the Republic of Georgia, and are just under the surface in places like Saudi Arabia and Iran. The security dimension is further implicated in that the vulnerability of supply creates the need to protect pipelines, refineries, and tanker sea-lanes around the world. Finally, it should not be lost on us, especially in the midst of trying to extract ourselves from a second Gulf War, that throughout the last century, oil has been a contributing cause of military conflict. A clean-energy economy would break the world's dependence on a small set of security-fragile regimes, stop funding nondemocratic states, free up resources for other forms of protection, and allow us to deploy money and human energy toward safer and more productive uses.

By pointing out the benefits of responding to climate change, I do not wish to minimize the genuine dangers or be Pollyanna about the challenges of transitioning to a post–fossil fuel world. Rather, I want to emphasize the silver lining in our predicament and feature it. Ironically, we can see this in the end of nature arguments. By turning attention away from biophysical constraints and reminding us of how much our choices are about the world we want to create rather than the one we feel we need somehow to align ourselves with, we are given greater freedom to explore the social advantages of one set of affairs over another. Biophysical thresholds call out calamity; social benefits promise well-being. Environmentalists would do well to emphasize the latter.

A second step involves establishing a new relationship to energy. When we see ourselves as the main culprits of environmental degradation and work to protect the natural world from our forays, we tend to envision energy sourcing as a matter of crossing a boundary, securing energy resources, and coming back to burn or otherwise utilize them. We see our job, in other words, as a matter of robbing nature to power our lives, rather than seeing ourselves as part and parcel of our energy sources.

To those enticed by the mastery narrative, the earth's crust is basically a large rock, and its atmosphere is an inert container positioned to receive our waste. Coal, natural gas, and petroleum are thus dead substances that are useless aside from our employment. In fact, we don't even see them if we don't extract them from under the earth's surface. The air is likewise an invisible realm that stretches around the earth with no special status aside from its ability to provide oxygen and absorb carbon dioxide and other pollutants. Seen as such, there is nothing wrong with an energy system premised on extraction and emissions. The dynamics involved take place external to our lives.

In the case of energy, a postnature orientation calls on us not to steal energy from the earth and emit waste into the air, water, and soil as if these were receptacles separate from human life but instead to find ways of harnessing the earth's energies so that our lives and the dynamics of the nonhuman world intertwine. When we mine and burn fossil fuels we use up nature's capital. In contrast, when we harness the wind, sun, ocean waves, rivers, and geothermal activity, we tap into inexhaustible forces. The difference is profound.

Drawing attention to using renewable energy sources is, of course, nothing new, and focusing on them is certainly not unique to a postnature orientation. But such an orientation draws our attention more tightly to them. A postnature sensibility does not scribe a line between us and the nonhuman world, and call on us simply to stay on our side of the divide, nor does it necessarily counsel excessive restraint when it comes to interacting with the nonhuman world. The nonhuman world is not off-limits. Rather, it has, like humans, inexhaustible energy resources. The key is an orientation to work with, rather than against, these—to partner our lives with, rather than lord over, them. Moreover, a postnature sensibility reminds us that we are already fashioning the nonhuman world, and celebrates this. Building wind turbines, solar collectors, hydroelectric systems, and so forth requires tremendous ecological intervention on the part of humans. The intervention itself is not a bad thing; it only becomes that way when we hold ourselves wholly separate from the nonhuman world or when our interventions exploit the earth. Harnessing the earth's energy makes sense in a postnature world.

A third step involves expanding the meaning and practice of adaptation. For close to a decade there has been debate about the relative merits of mitigation versus adaptation. Through much of this, many environmentalists rejected adaptation out of hand because they felt that we should first do everything we can to prevent climate change before essentially throwing in the towel and starting to prepare for it. Despite the insights and concerns of adaptation critics, the debate has now pretty much died down. It is not that one side's position persuaded the other but rather that the realities of climate change have eclipsed the debate's meaning. Climate change has begun. For better or worse, people have started to adapt to it. Municipalities, states, and international organizations are currently taking measures to prepare for climate change—building higher seawalls, constructing stronger levees, relocating populations, rechanneling inland waterways, and building satellite information networks to track and warn against imminent climate dangers. In fact, in 2007, the IPCC itself recommended a broad strategy for adaptation and preparedness, recognizing that to do otherwise would be morally and pragmatically inexcusable.

A postnature orientation celebrates moving beyond the mitigation-adaptation debate, but adds a third element: a concern with suffering. Mitigation and adaptation aim to lessen the amount of pain associated with climate change. Mitigation does this by trying to prevent the consequences of climate change; adaptation does it by taking measures to adjust our material lives so that we can continue much of our lifestyles with only minimal disruption. Neither, however, focuses specifically on working with the pain that is already visiting victims of climate change and that promises to spread as we enter advanced stages of climate change.

There is a subtle yet important difference between pain and suffering. Pain is part of the world. Everything living experiences times of hardship, deprivation, illness, and loss, and to the degree that beings have nervous systems, these are experienced as bodily or emotionally distressing occurrences. Suffering, in contrast, is less about physical or emotional hurt than about the fear and resistance associated with pain. Much of our suffering has to do with worrying, strategizing, and generally distressing about the experience of pain itself. We tell ourselves stories about its depth, meaning, future trajectory, and significance, and find ourselves anxious, alarmed, and at times even panicked by such narratives. Make no mistake, there will be much pain associated with climate change. People's lives will be upended, food and water scarcity will certainly contribute to deprivation, and widespread illnesses will undermine health. Surrounding all of this will be the stories we tell ourselves about the pain—and the stories we will tell ourselves as we anticipate the pain—and these will have profound implications for how we live together in a greenhouse age. We cannot, of course, avoid telling ourselves and living out various narratives. Yet some are more productive and induce less suffering than others. Living in a postnature age requires choosing those that are most promising.

There is no easy way to develop such narratives, but focusing on the process is important. We can do this by conducting dialogues across various disciplines, including not only traditional fields like medicine, psychology, and sociology but also the so-called wisdom traditions that have long reflected on the meaning and tried to cultivate practices for responding to the world's pain. We need to find common vocabularies for ways of enduring physical and emotional dislocation, experiencing grief, and expressing anguish. We should conduct widespread public discussions about shared ways of working with pain as a strategy for assisting in our collective ability to live through the greenhouse age in a humane manner. We may not know the "right" way to experience pain but we certainly know destructive ways to do so. Centuries ago, Malthus gave us one vision of how the pain of environmental degradation might be experienced—through racism, conflict, violence, and despair. We can do better. We can humanely live through our days. The key to doing so, though, is to begin meaningful public discussions through which we can share insights, practices, and collective commitments to assisting each other in the challenge of living through the inevitably painful dimensions of climate change.

What makes this postnature? Why is such a suggestion meaningful to a postnature age? A focus on suffering does not invoke either the dream of naturalism or mastery. Neither does it take a stand in terms of whether nature truly exists as a knowable realm. Rather, it confronts the challenge of humanity's fate and that of all living creatures as we move deeper and faster into a climate-changed world. It understands that people will be trying both to alter the earth and restrict or curtail human behavior to avoid climate change dangers. It does not take a stand on either of these but instead focuses on building connectivity between people through a sustained dialogue on the meaning and strategies for humanely experiencing together the inevitabilities of climate change. This is not old-time environmentalism or technophile mastery. It is "post" the twin urges that have fueled the polarized sides of environmental politics as well as the philosophical sensibilities that have animated and torn us for centuries.

These three steps toward a postnature climate age share a common orientation toward conventional environmental politics. They emerge out of a perspective that sees neither the dream of naturalism or mastery particularly appropriate for our postnature times. We cannot respond to climate change by simply getting out of the greenhouse gas business. The huge amount of carbon dioxide we have already emitted into the atmosphere has committed us to a hotter, biologically depleted, and unpredictable-weather world. The dream of naturalism suggests that we can best align ourselves with nature by shrinking our ecological footprint and letting nature—which in this case means the atmosphere—be. This is not a viable option, however. Neither is the alternative: the implementation of a full-court press toward mastery. Mastery brought us climate change; we should be especially wary of employing it to get us out of it. The ideas outlined above strive to strike a middle path through the two dreams.

Conclusion

None of what I've discussed above may strike the reader as profound. The ideas are not fundamental shifts in strategy but rather particular emphases on certain policy trajectories. They are exercises in extending insights from a postnature orientation toward climate change. They do not, obviously, solve our climate change dilemmas but instead, it is hoped, deepen our appreciation for the depth of the challenge and steer us in certain directions rather than others. Focusing on the benefits of transitioning to a clean-energy economy, choosing only renewable energy sources that require harnessing the earth's power rather than using stored-energy capital, and developing public debate about strategies for enduring the inevitable pain that will accompany climate change are not specific lines of attack or particular tactics; they are general orientations with which to develop specific strategies.

My main point is that our climate change politics can no longer be tied to biophysical imperatives but rather must emerge from collective decisions about the kind of future we want. There is nothing inherent in the nature of the universe that demands one set of climate conditions or strategies over others. In every case, some people and nonhuman creatures will suffer, and some landscapes and ecosystems will change. How many and which people will suffer most? To what degree will ecosystems change? How hot are we willing to let things get? How much climate disruption are we willing to endure? These kinds of questions are open to different answers. Nature itself (that is, climate itself) provides no single response. The kind of answers we end up supplying will not, of course, be purely technical. They will and should involve ethical, aesthetic, pragmatic, and spiritual sensibilities. Such sensibilities are all we are left with in a world in which climate itself is already inflected with and increasingly determined by human beings.

A postnature climate orientation, like the one I outlined in the previous chapter with regard to wilderness, concerns itself with boundaries. In a postnature world, boundaries make less sense, but they are not completely insignificant. We must soften the boundaries between humans and the wider world of which we are a part while concomitantly recognizing the analytic benefits they provide. Given this, the steps toward a postnature climate age I have outlined try to avoid the distinction between humans and nature. They try to think about responses to climate change that take seriously the interdependencies between the human and nonhuman worlds, and not play a game of sequestering nature from human intervention.

Climate change represents the most challenging environmental dilemma on the horizon. Responding to it requires, at a minimum, revamping the world's energy systems and building resilient communities that can assist each other through severe climate disruption. As many others have made clear, there are no silver bullets in responding to the climate challenge; there are only sets of choices—of technologies and practices—which we must work out together.[23] At the heart of making those choices is, as the word "responsibility" makes clear, our *ability* to respond. By facing up to the realities of a postnature world, and cultivating ways of shifting collective life in meaningful and effective directions within those realities, we can hone our response abilities. We will not solve our climate woes by doing this but instead will be able to address them with as much insight and humanity as possible. This may seem like small consolation. But these days, I'm unsure what other responses are honest and engaged enough to be more relevant.

"Space between Two Walls," Les Baux de Provence, France

8

Being an Environmentalist: Decisive Uncertainty and the Future of American Environmentalism

Take your well-disciplined strengths
and stretch them between two opposing poles.
Because inside human beings
is where God learns.
—Rainer Maria Rilke, "Just as the Winged Energy of Delight"

For decades, American environmentalists have seen the natural world under attack and taken up the charge of defending it from the exploits of humanity. Efforts to combat climate change, ozone depletion, loss of biological diversity, fresh water scarcity, and the like all reflect this sensibility. They express a commitment not to sit by and let humanity undermine the life-support system of the planet or otherwise degrade the natural world but rather to stand with nature and defend it from ecologically damaging assaults.

As this book has documented, American environmentalism's long-standing attraction to nature is coming undone. Nature is no longer an independent realm separate from human beings but instead part and parcel with the human world. Empirically, humans have erased the boundary between the human and nonhuman spheres; conceptually, we have come to see through nature's independent status. As a result, environmentalists are now living in and need to adjust their mission to a new world. We must find fresh ways of working on behalf of environmental protection in the absence of a stable, well-understood notion of nature, and within a political landscape in which debate must go beyond the merits of nature versus humans. Such a challenge requires not only altering agendas, building new alliances, and engaging in the political process in different and more creative ways. It also involves, more fundamentally, being a new kind of environmentalist. Without nature around to orient one's work and life, American environmentalists must develop new understandings of their own and humanity's place on earth, and translate that understanding into political practice. Such is the environmentalist task in a postnature world.

Throughout the preceding chapters, I have suggested that a new mode of environmentalism can emerge through what I have been calling a middle path. A middle path is a route into the postnature world that leaves behind the compelling attraction of either the dream of naturalism or mastery, and embraces a politics of ambiguity. In this final chapter, I describe the outlines of this middle path, and explain its promise for American environmentalism.

Political Ambiguity

Few see ambiguity as a virtue. In the midst of complexity, most of us want to find something secure to trust in, something around which to base our understandings and on which to rest our sights. This is also the case when it comes to environmentalism and explains why so many environmentalists, especially in the United States, have counseled staying away from questions about the nature of nature. Many fear that reflecting on changing understandings of nature will drag the movement into a kind of never-ending navel-gazing that will only undermine confidence and passion. The last thing the movement needs right now, they would argue, is doubt and indecision. In the face of today's ecological degradation and especially the gathering storm of dangers associated with climate change, the movement is looking for decisive action, and ambiguity seems anathema to clear thinking and determination.

As I have been trying to demonstrate throughout this book, ambiguity need not muck things up by dimming environmentalists' sights but can actually provide insight and the kind of perspicuity longed for in these difficult times. It can, in fact, supply confidence and direction. It can do so to the degree that it stretches environmentalists across the tensions that inflect environmental decisions in ways that demand integrity.

All of us live contradictions. Few of us walk the talk. The problem is that we are forced to hide our contradictions for fear that they will undermine our advocacy or compromise our public persona. A postnature age requires us to acknowledge this, and seek to move beyond it by enlarging ourselves to include the cross-cutting pressures on our environmental lives and forge direction as a result. In the preceding chapters, I suggested that a way to do this is to situate oneself at the interface between the broken dreams of naturalism and mastery, and approach environmental challenges stretched uncomfortably across the divide.

As the chapters on wilderness protection and climate change indicate, environmentalists may be attracted to the idea of harmonizing with nature, and be deeply devoted to preserving the organic quality of the nonhuman world and desisting interventions into so-called natural processes. But this does not mean that we reject all intervention or rebuff tendencies toward mastering nature. As shown in chapter 6, protecting wildlands and wildlife, while resonant with the dream of naturalism, takes a tremendous amount of human energy, including using some of the most sophisticated technologies and management techniques. Leaving nature alone as a wilderness protection strategy—reservation ecology—is simply not an option in our humanized world. Likewise, as discussed in chapter 7, leaving the carbon cycle alone so that it can do its "thing" is not a viable strategy for combating climate change. We have crossed too many thresholds to get out of the greenhouse gas business. We are now influencing global climate, and we better recognize this and direct our actions accordingly. In both cases, we must concede how the imperatives of the day force us to span the naturalism-mastery divide and arrive at a middle way that carefully yet confidently operates in a postnature age.

Key to doing this is realizing that our orientations are not compromises due to contemporary exigencies but rather inevitable paradoxes that give expression to deeper impulses within each of us. While environmentalists tend implicitly to subscribe to the dream of naturalism and skeptics tend to subscribe to the dream of mastery, in truth there is a little bit of both impulses in each of us. Personally, I love the woods *and* the city; I enjoy hiking *and* reading the newspaper. When I am sick, I count on my body to heal itself *and* take synthetic medicines to outsmart and override my body's dynamics. Politically, environmentalists engage in the same kind of jujitsu. For instance, while we worry about climate change, and advocate reducing our use of cars, electronics, and other technologies to address it, we also depend on and employ extensive technologies to understand the character of climate change and forge paths toward responding to it. (Many of us also fly around on planes so we can lecture others about how to reduce their carbon footprint.) Likewise, we often counsel letting rivers run free in the sense of dismantling dams, reducing water pollution, and restoring river watershed features. Nonetheless, we utilize sophisticated water-sampling techniques, ladders to assist salmon crossing barriers, and bulldozers to affect this. Our actions reveal that we are quite accustomed to both lording over *and* harmonizing ourselves with the nonhuman world. The problem is that we are not confident in doing so. We can gain confidence by seeing what we used to assume was a conflict as simply the experience of living in a complex world in which the old standards of value and political engagement still murmur in the background, but no longer provide the secure insight they once did. A postnature age welcomes us into this world rather than pretending otherwise.

The Middle Path

To say that ambiguity provides a route toward confidence and integrity in a postnature world may seem to be just another paradox. How can ambiguity supply any sense of assurance and support conviction? Isn't ambiguity the opposite of confidence? It all depends, of course, on what one means by confidence, and how deeply one can experience it.

The dual dreams of naturalism and mastery have long offered their respective camps philosophical places of confidence. They have provided coherent worldviews and prescriptions for practice. Armed with such dreams, environmentalists and their critics enter debates not at square one—as if coming to environmental affairs each time anew—but with developed tools of analysis and plans of action. This enables them to categorize data, filter information, position themselves quickly within debates, and formulate decisive action. It also, however, blinds them to facts, contexts, and forms of interpretation that fall outside their purview, and creates an undue sense of smugness about one's understanding of the world. The dual dreams suggest that we already understand the world: we know what is most important to know, and need only figure out how our categories of understanding can be applied in given instances. This provides a type of confidence that can make one feel secure and assured, but to the degree that it does so by shutting out the unfamiliar, it gives one a shallow or even false sense of confidence. It offers conviction at the price of narrowness. With their deep commitments, the dreams lock out novelty and surprise. They bar their adherents from what those who value wildness have long considered paramount: the novel, spontaneous, and unbidden. And without this—without an openness to the different, strange, and untried—deep down it is hard to feel self-assured. In a paradoxical sense, confidence comes not from knowing everything and being able to control our experience but rather from knowing that we do not know everything, and nonetheless finding ways to live meaningfully and work on behalf of life.

Ambiguity finds its spirit in wildness. We are most indefinite when we are not in control and when the world appears to us uncontrollable. At such moments we know, with certainty, that we do not and cannot know everything. This becomes a virtue to the degree that we embrace our unknowing. That is, we gain confidence by understanding the limits of our understanding. *This* is the key to preserving wildness in a postnature age. To appropriate Thoreau, one could say that being torn and unsure about the way of things *is* the "preservation of the world."

But what does *this* really mean? What is a middle-path environmentalism rooted in ambiguity?

The middle path of ambiguity involves cultivating a decentered environmental politics. For too long, nature and humanity have provided candidates for the center of our environmental concerns. They have served as fixed foci around which we organize our thinking and action. As political scientist Kerry Whiteside puts it, a center is a "unique thing endowed with certain properties allowing it to illuminate the value of other things. The goodness of those things becomes a function of their proximity to the 'center.'"[1] In other words, the dual centers of nature and humanity create ethical systems that assign value based on how close things orbit either the natural or human worlds. These centers do not disappear in a postnature age but rather become merely nodes or frames of reference that exert less gravitational pull on our thinking even if they murmur their ancient refrains. A decentered environmental politics means hearing and acknowledging their hum without being tied to it. It suggests that we can enter environmental debates with our eyes wide open—aware of our complex impulses and the theological poles to which we are used to be being drawn.

What happens when we do so? The first thing is that we realize that environmental questions can no longer be a debate about the relative importance of nature *or* humans. They can no longer be framed as trade-offs between what is best for humanity *versus* nature. Instead, they are about humans *and* nature, and the mutually constitutive dynamic between them. Because a strict boundary between the human and nonhuman worlds no longer exists, we need to recognize that life has only one fate. Yes, individuals and specific entities may have distinct destinies, but the humanization of the earth has set everything on an interdependent path in which human action and nonhuman realities share a common experiential and evolutionary trajectory.

Many American environmentalists worked hard in the 1970s and 1980s to protect endangered species and ecosystem hot spots in the developing world. They sent scientists to study the mating patterns, dietary requirements, and environmental conditions of various plants and animals, and policy experts to devise ways of cordoning off ecologically rich areas and threatened species from poachers, habitat destruction, and other dangers. In many cases, such environmentalists were successful in setting up nature preserves, sanctuaries, or other forms of legal protection. Despite such efforts, many environmentalists eventually realized that they were fighting a losing battle. Endangered animals were still disappearing, lands were still being encroached on, and ecosystems were still being destroyed. There are various explanations for environmentalist failures, but one factor is surely environmentalists' overconcern for the nonhuman world at the expense of the human one.

By circumscribing and patrolling preserves and sanctuaries, environmentalists often ignored the people who were living inside or close to such boundaries. As a result, environmentalists failed to address the incentives that people have to destroy habitat, poach, and so forth. Many environmentalists fortunately have learned their lesson. They recognize that you cannot protect wildlife or wildlands without paying attention to the well-being of people. Such understanding has spurred various efforts to include residents in decisions and practices associated with

nature preserves—for example, incorporating resident populations in local ecotourism, or making sure that revenue generated locally from preserving wildlife stays in the community
rather than being exported to the capital or simply outside the
area. To put it differently, it has led environmentalists to care
about human needs along with the environment.

There is much good that has come out of such learning.
One problem with it, though, is its strategic character. Environmentalists concerned with plants and animals in the developing world *have* to pay attention to human needs if they want to
succeed in their wildlife protection efforts. They focus, in other
words, on humans as a concession to achieve other goals. This
is not to say that environmentalists do not care about humans
but merely to note that the reason many of them have had to
go into the human needs business is out of concern for the
nonhuman world. Plants and animals are the focus; humans
are the means.

We see a similar story in many environmentalist actions. In
the parlance of environmental discourse, focusing on humans
is a type of "brown" environmentalism. For too long, environmentalists have concerned themselves with "green" issues: the
well-being of plants and animals, the preservation of certain
landscapes, and the protection of ecologically rich or otherwise aesthetically pleasing areas. Over the years, they have been
dragged into brown issues: urban health concerns, sustainable
development, the siting of toxic waste dumps, incinerators, nuclear power plants, and the differential experience of environmental harm. In many ways they have come to these campaigns
kicking and screaming insofar as such issues have taken environmentalists far from their conventional mission. The result is
that in many environmentalist circles, brown issues are channels toward green concerns. They represent the necessary work
that needs to be done to protect nature in an effective way.

This is, of course, changing. Today, the environmental
movement has a strong wing focused on environmental jus-

tice, sustainable cities, green technologies, and other elements of sustainability understood in the broadest sense. Recognizing that we live in a postnature age further emboldens these efforts by inviting us to rethink the relationship between means and ends. Without nature around to serve as a good toward which to orient our campaigns or a goal in which human-needs work is merely an instrument, we must see humans and nature as fundamentally linked. As theologian and environmental thinker Thomas Berry observes, "We now in large measure determine the earth process that once determined us. In a more integral way we could say that the earth that controlled itself directly in the former period now to an extensive degree controls itself through us."[2] This means that what we think, imagine, construct, and organize, we do for ourselves and all creatures. Conversely, it also means that our thoughts, fantasies, and constructions are not self-originating but instead develop in relationship to the more-than-human world of which they participate. The nonhuman world works through us to the degree that we are the main agents of planetary change, but there is also something beyond us. Just because the world has become humanized—with a human signature everywhere—does not mean that there is nothing except humans. A bio-engineered tree may express human design, yet it still is not a human being, and as mentioned, a genetically modified mouse may be programmed to develop cancer, but it is still a mouse and not a person. In a similar vein, while human beings today may have animal organs and even possibly altered DNA, this does not make them nonhuman. Political theorist Francis Fukuyama and others may refer to such people as "posthuman," but this does not get rid of the fact that these beings are not orchids, frogs, or sandstone.[3] They are not, to be sure, the kind of human beings that we have long conceptualized—with an essential and stable set of bodily and cognitive features (expressing a given human nature). But this doesn't subtract from their identity as humans nevertheless. The most souped-up human—

equipped with artificial limbs, a monkey's heart, and changed DNA—is still human. And yet it is not simply that. All living and nonliving entities on earth are a mélange. We are so intermixed and mutually constituting that although we are different entities, one cannot disaggregate the human and the nonhuman, nor imagine their fates as separate.

This goes not simply for day-to-day experience but also for our evolutionary track. In a postnature age, we are on a coevolutionary adventure in which humans and the more-than-human world forge a common trajectory. One sees this with climate change. Humanity's evolutionary trajectory, which includes industrialization and the widespread use of fossil fuels, is altering nature's evolution in the form of climate change. The earth's systems are changing in response to human activities. The dynamic continues, however, insofar as climate change is now altering human consciousness and promises, if we are lucky, to set humans on a different industrial course. Of course I am using the notion of evolution here in a liberal manner, but it should be clear that the coevolutionary dynamic is at work in even the strict sense of evolutionary change. Whatever humans and nonhumans look like in the distant future will be a result of the interactions between them or, put differently, their bonded identities. *Human/nature* is what everything is about these days—a hybrid world in which there is no such thing as either humans or nature per se but rather an amalgamation or fusion of the two. Put differently, humans may be in the earth's evolutionary driver seat, but our actions are not self-originating. They emerge out of hybridity.

To realize that we are on a coevolutionary journey calls on environmentalists not to see human needs and nature's needs as separate goals but instead one and the same. Concentrating on human well-being is not a concession made in the enterprise of environmental protection; it is part and parcel of a broader environmentalist project.

The Middle Way of Bright Green Politics

Another consequence of a postnature age for environmentalists has to do with our appreciation of human ingenuity, technological prowess, and other dimensions of the mastery narrative that we have long resisted. American environmentalism has long sought to erect boundaries between humans and nature to protect the latter from the former. This has involved policing nature preserves and preaching restraint when it comes to using natural resources. A postnature age questions this orientation, and invites us to welcome and give greater significance to a different approach espoused by an emerging wing within the movement.

After years of urging people to sacrifice their material desires in the service of environmental protection, a new crop of environmentalists is claiming that humans need not hold back, restrict, reduce, or otherwise diminish our ecological presence but rather embrace that part of the human spirit that seeks innovation and enjoys extending our imprint across the earth. We see this type of environmentalism being expressed in efforts to design our way out of ecological woes, technologically surmount issues like climate change and peak oil, and leave the whole idea of nature behind as an anachronistic category of environmental analysis.[4] This type of environmentalism is most focused on building sustainable cities, improving industrial metabolism, investing in mass transit, outfitting the world with solar panels, and a whole host of technically demanding challenges. This breed of environmentalist tends to be upbeat about our prospects and tempers a quiet dream of naturalism with a strong attraction to technical prowess. To distinguish this orientation from earlier movement expressions, some have come to describe it as "bright" green politics.

With faith in technology and a "can do" sensibility, bright greens certainly flirt with, if not get in bed with, the dreamers of mastery. Ordinarily this would concern environmentalists, and historically the more conventional wing of the movement has worked to mute Prometheanism from within (witness the many criticisms of the bright greens by their more traditional brethren).[5] A postnature age signals that earlier antagonisms might best be left to the past, and that we can now incorporate, to a degree, the Promethean voice within the movement much more comfortably. For without the "god" of humanity behind the dream of mastery, environmentalists can gesture toward such a dream without worrying about abandoning established and cherished principles. One does not leave nature behind when one embraces technology, ingenuity, or humanity's ability to control nature. Rather, one acts with eyes open to the realities of hybridity; one embraces the human/nature mélange in which we live. This enables us to develop practices that privilege neither humans nor nature but instead respect the blend of the two. For example, it would support efforts at adaptive management, environmental design, sustainable urbanism, and precautionary politics.[6]

At the heart of adaptive management, environmental design, precautionary politics, and other postnature orientations is the type of decentered orientation mentioned above, and a kind of stewardship mentality that seeks to cultivate the human/nature world toward health and sustainability rather than working on only one side of what we have heretofore taken to be a divide. One of the problems with the dual dreams of naturalism and mastery is that they both assume particular stances toward wildness. The dream of naturalism assumes that the way to reveal wildness is to minimize human presence. The dream of mastery seeks to stamp out wildness or, more accurately, sees wildness simply as an unwitting victim of our efforts to control the world. Postnature approaches enable us to go beyond the either/or frame by seeing us as caretakers and cultivators of the wildness both within ourselves and the wider world of which we are a part.

In his perceptive book, *The Case against Perfection*, political theorist Michael Sandel asks what is wrong with bioengineering our progeny.[7] He wants to know what is it that strikes many of us as disturbing when we imagine the possibility of genetically altering our children. For instance, he examines questions of risk: What if things don't work out the way we imagine? What if, instead of blue eyes, our kids have no eyes? There are certainly huge dangers in altering human DNA, and we should thus tread lightly. Additionally, Sandel looks at certain principles that we have long valued like human liberty and fairness, and worries about their fate as we contemplate genetically modifying our species. Will genetically altered people be genuinely autonomous or simply subject to the genetic instructions imparted by their genetic architects? Likewise, if genetic manipulation is available only to the wealthy—which, at least initially, would be the case—might it breed an aristocratic race in which only certain people will be able to afford heightened musical talent, concentration, athleticism, and conventional notions of beauty? Much of Sandel's book reviews and ponders such concerns. Sandel concludes, however, that our deepest hesitation toward genetically engineered children goes beyond functional or social consequences. Rather, it has to do with our sensitivity toward what he calls the *giftedness* of life. Sandel explains that our hesitation comes from an appreciation for the unpremeditated or unbidden character of our children. Our children arrive, in other words, unscripted by us or anyone else. As such, they present us with sheer novelty and unpredictability. For Sandel, to genetically modify our progeny is to embrace the mastery ideal, and snuff out the surprise, wildness, or otherness that birth manifests. Our children bring new worlds into our lives and human experience in general, and this, Sandel argues, is priceless.

Sandel's reflections provide a metaphor for thinking about a postnature middle way. In a postnature age, although there is no nature to nature or humanity, this doesn't bleach out the giftedness of our experience. The tension between the dreams of naturalism and mastery ensures as much. We may genetically manipulate human nature, but this does not mean that we completely control human experience. As well, we may dominate the nonhuman world with our power, yet this does not mean that we get rid of it altogether. The wildness of the world and our selves may be muted in an age when humans continually delve into and seek to manipulate human experience as well as the workings of the nonhuman world. But it is not extinguished. Rather, it is part of the mélange in which we now operate. It is folded into the broader human/nature reality. The challenge for environmentalism in a postnature age is to keep this alive and present.

For Sandel, our attitude toward the giftedness of life should be analogous to our approach to parenting. On the one hand, we are responsible for enabling our children to flourish in the world. This involves teaching, materially supporting, and orienting them to be the kinds of people we think would be happy and contribute meaningfully to society. On the other hand, we must respect their givenness. We must appreciate them as gifts that we accept as they come. Our children, in other words, are not instruments of our ambition or products of our will. They have their own ways that we should cultivate and allow them to express. In short, we should behold our children as well as mold them.

Our dual orientation to parenting is instructive for environmentalism in a postnature age. Like children, the human/nature condition of our age is neither an object of our manipulation nor something that can simply will itself aside from our participation. We thus need to bring both our designing selves to the world of which we are a part, and a sense of humility about the real and preferred limits of our interventions. This means we take an active role in shaping our world, but also respect the wildness that harbors within that world, and let it express itself and shape us. We need, in essence, two types of approaches to our environmental challenges: a sense of acceptance for the given quality of things, and a transforming attitude that seeks the well-being of all things.

The Middle Path in Dark Times

For too long, environmentalism has been about doom and gloom. The modern environmental movement of the 1960s and 1970s peddled in apocalyptic scenarios that aimed to scare people to action, and much of the movement still uses fear for motivation. There is, of course, much to fear these days. Today, one in four mammals face extinction; everything frozen on the earth is melting due to climate change; water scarcity is endangering millions of people and untold species; and forests are disappearing at alarming rates. If one isn't gloomy in this world, one might say that he or she simply hasn't been paying attention.

But as mentioned in the discussion about climate change, doom and gloom has its drawbacks. Fear may motivate some people yet it turns off many others. Moreover, it inspires people to advance specific warnings that end up discrediting the movement if they fail to materialize. Furthermore, fear pushes people toward desperate acts that can be inappropriate and dangerous in their own right (for example, the radical geoengineering schemes I discussed in the previous chapter). Fear has its limitations at the personal level as well. As Leopold tells us, it is hard to live in a "world of wounds."[8] It is difficult to hold the sorrow that many of us feel in witnessing the desecration of our world, and the injustices and ugliness associated with this. Such sorrow breeds pessimism and despair, and this can often increase a sense of despondency and depression.

Environmentalism's doom and gloom has further ramifications. Let's face it: many of us as environmentalists are a pretty sorry lot to hang out with. We walk around angry at our fellow citizens for the ways they squander resources and fail to appreciate the preciousness of the earth, and voice a constant complaint about global and regional inequities when it comes to the distribution of the earth's bounty. We look around at our societies and find little to admire, and spend our days under a dark cloud waiting, like Chicken Little, to be proved right. We appear as fearmongers or disgruntled folk unsatisfied with all that society has to offer, and hang our heads low and talk to ourselves and others about how things keep getting worse. Such a demeanor is not only a sorry way to feel; it is also an immense turnoff to those we wish to join our cause.

In addition to muttering complaints to ourselves and others, we also experience much guilt about our own lives. Ensconced in systems that offer few choices to live ecologically sustainable lives, we walk around feeling guilty all the time because so many of our actions fall short of our ecological ideals. Many of us feel remorseful when we travel in fossil fuel machines, eat foods that we know have been grown in ways that deplete the earth's fertile soil, or overuse water in places where it is scarce. We have created environmental thought police within our own heads that constantly monitor our lives and criticize us for our multiple transgressions.

Being an environmentalist becomes even harder as we realize that the battles we engage in transcend our lives. Environmental issues are not puzzles in search of solutions but rather perennial challenges that each generation must face anew. For all intents and purposes, climate change will be with us forever. As long as there are fossil fuels, anthropogenic climate change is possible. The same goes for threats to wilderness. As long as there are reasons to use natural resources and create waste, wilderness will be vulnerable. Environmentalism is a matter of eternal vigilance. It is a full-time job that requires of us a lifetime of concern and commitment to goals that transcend our own lifetimes, and those of our children and grandchildren. If environmentalism is all doom and gloom, this makes for a sad, guilt-ridden, and (often) angry vocation. Such an orientation has to be hard to sustain, if not fundamentally damaging to our own lives. Moreover, if environmentalism is all doom and gloom, it makes it extremely difficult to win converts. Few people get excited by the prospect of worrying about things that resist easy solution and promise to be around well beyond one's individual life. It is so much easier simply to withdraw and go about our lives blind to the threats to environmental well-being. "Oh well!"

Understanding the postnature world relieves environmentalism, at least a bit, of all the worry. It helps us see that environmental protection is not a battle in defense of nature against a growing and increasingly powerful humanity. Rather, it involves safeguarding the world of humans and nature, or more accurately, the hybridity of which we are a part. Recognizing this should help insofar as we need not see every ecologically insensitive human act as another trouncing of nature and thus a lost battle but instead as another turn in a world that continues to evolve. We may not like the change of direction, but with a postnature mind-set, we need not understand it simply as another nail in nature's coffin. Thresholds will be crossed certainly, but new realities will present themselves, and we need to maintain a quality of mind and a movement spirit that can find the courage as well as resources to act in these new realities in ways that will cultivate health, justice, and ecosystem well-being. Our ability to maintain such an orientation rests on appreciating our unknowing. It consists in recognizing that no matter how much the world may appear overlaid by humanity, deep mysteries abide both inside and outside ourselves whose wildness is crucial to maintaining our own sense of well-being along with that of the world.

The unknowing I am talking about is not, of course, ignorance. Rather, it involves understanding what can actually be cognitively grasped, and what is beyond the reach of conceptualization, systematization, and general comprehension. In a postnature world, it should be clear not only that we *cannot* know everything but in important ways we *need* not understand everything. The urge to comprehend everything shares much with the dream of mastery. The urge to know everything is related to the impulse to control our world. If we understand the inner workings of things, we can then manipulate all that is around us. Free from the dream of mastery, we can begin to focus not on knowing everything but rather on understanding what is important. A postnature environmentalism is about creating a livable world for all. These days we need to focus on what makes that possible, and how we and other living creatures can flourish in doing so. Traditional environmentalists may insist that creating a livable world involves reestablishing the boundary between humans and nature, and staffing the barricades. But this, as should be apparent, is not an option. What is an option is recognizing the collective fate of humanity/nature and then righting ourselves to the mysteries inherent in that mélange.

Environmentalism, especially its American variety, has traditionally been about protecting nature from humanity's onslaught. If we dig deep enough, however, we recognize that the idea of nature in this formulation has referred not strictly to empirical reality but has been used as a conceptual stand-in for the notion of *otherness*. American environmentalists have traditionally understood nature as that which is not human. It represents the self-willed world independent of conscious human influence. In this way it is the epitome of otherness. In a postnature age, much of this representation—in the form of wilderness areas, expansive landscapes, and species diversity and abundance—has been altered, often beyond recognition. But underneath the alteration, otherness still exists. To be sure, it is harder to see, access, and experience, yet getting in touch with it and trying to keep it not just alive but also present in our experience is central to a postnature environmentalism. Keeping otherness alive feeds our sense of excitement at coming to the edge of our knowledge and control, and serves as a strategy for making that edge the center of our politics. The more we honor otherness, the more we will seek its cultivation in ourselves and our world.

Our postnature condition calls on us to redefine environmentalism. As the divide between humans and nature disappears, we must realize that environmentalism as a *nature* movement is anachronistic. In its stead, we must develop a genuinely *environmental* movement. This means focusing attention on the world around and within us—as it is given and as it can be transformed. The given, in this case, is not merely the biophysical other-than-human sphere but the human one as well. We meet the given as a historical moment. Environmental politics is about bringing public power to bear on how we respond to that moment. We need to respond in ways that enhance rather than diminish life, and in ways sensitive to the long-standing values of justice, economic well-being, peace, and ecological sanity. Key to all of these is respect for and a desire to care about the other. Environmentalism has long used the concept of wildness to capture the other. The good news is that even in a postnature age, wildness is alive and able to be shared. Sensitizing ourselves to wildness is the future of American environmentalism.

Notes

Chapter 1

1. Aldo Leopold, *A Sand County Almanac and Sketches Here and There* (New York: Oxford University Press, 1989), vii.

2. Ibid., 197.

3. Henry David Thoreau, *Walking: A Little Book of Wisdom* (New York: HarperCollins, 1994), 19.

4. Thoreau, *Walden*, 87.

5. René Descartes, *Discourse on Method and Meditations on First Philosophy*, trans. Donald A. Cress, 4th ed. (Indianapolis: Hackett, 1999), 35.

6. Bill McKibben, *The End of Nature* (New York: Random House, 1989).

7. See, for example, Luc Ferry, *What Is the Good Life?* trans. Lydia G. Cochrane (Chicago: University of Chicago Press, 2005); Kerry Whiteside, *Divided Natures: French Contributions to Political Ecology* (Cambridge, MA: MIT Press, 2002); Kate Soper, *What Is Nature? Culture, Politics, and the Non-Human* (London: Blackwell, 1995); R. Bruce Hull, *Infinite Nature* (Chicago: University of Chicago Press, 2006); Timothy Luke, *Ecocritique: Contesting the Politics of Nature, Economy, and Culture* (Minneapolis: University of Minnesota Press, 1997); Frank Fischer and Maarten Hajer, eds., *Living with Nature: Environmental Politics as Cultural Discourse* (New York: Oxford University Press, 1999).

8. William Cronon, ed., *Uncommon Ground: Rethinking the Human Place in Nature* (New York: W. W. Norton, 1996).

9. William Cronon, "Introduction: In Search of Nature," in William Cronon, ed. *Uncommon Ground*, 21.

10. Carl Pope, *Strategic Ignorance: Why the Bush Administration Is Recklessly Destroying a Century of Environmental Progress* (San Francisco: Sierra Club Books, 2004); Albert Gore, *The Assault on Reason* (New York: Penguin Press, 2007).

11. David Orr, *The Nature of Design: Ecology, Culture, and Human Intention* (New York: Oxford University Press, 2002); Gary Snyder, "Is Nature Real?" in *Wild Earth: Wild Ideas for a World Out of Balance*, ed. Tom Butler (Minneapolis: Milkweed Editions, 2002); Edward O. Wilson, *The Creation: An Appeal to Save Life on Earth* (New York: W. W. Norton, 2006).

12. John Stuart Mill, *Three Essays on Religion* (Amherst, NY: Prometheus Books, 1998), 5.

13. See, for example, Matthew Parris, "Nature Superior to Man? What Green Twaddle," *TimesOnline*, August 30, 2008, available at <http://www.timesonline.co.uk/tol/comment/columnists/matthew_parris/article4636286.ece>.

14. Simon Young, *Designer Evolution: A Transhumanist Manifesto* (New York: Prometheus Books, 2006); Bjørn Lomborg, *The Skeptical Environmentalist: Measuring the Real State of the World* (Cambridge: Cambridge University Press, 2001).

15. Daniel Botkin, *Discordant Harmonies: A New Ecology for the Twenty-first Century* (New York: Oxford University Press, 1992), 193.

16. Quoted in Walter Truett Anderson, *Evolution Isn't What It Used to Be* (San Francisco: W. H. Freeman and Company, 1996), 145.

17. Holmes Rolston III, "The Wilderness Idea Reaffirmed," in *The Great New Wilderness Debate*, ed. J. Baird Callicott and Michael Nelson (Athens: University of Georgia Press, 1998); Donald Waller, "Wilderness Redux," *Wild Earth* 6, no. 4 (1996–1997): 36–45; Reed F. Noss, "Wilderness—Now More Than Ever: A Response to Callicott," in *Wild Earth: Wild Ideas for a World Out of Balance*, ed. Tom Butler (Minneapolis: Milkweed Editions, 2002).

18. David W. Ehrenfeld, *The Arrogance of Humanism* (New York: Oxford University Press, 1981).

19. Albert Borgmann, "The Nature of Reality and the Reality of Nature," in *Reinventing Nature? Responses to Postmodern Deconstruc-*

tion, ed. Michael E. Soule and Gary Lease (Washington, DC: Island Press, 1995), 176.

20. Alan Sokal and Jean Bricmont, *Fashionable Nonsense: Postmodern Intellectuals' Abuse of Science*, 1st ed. (New York: Picador, 1999).

21. Anderson, *Evolution Isn't What It Used to Be*, 175. See, generally, Walter Truett Anderson, *To Govern Evolution: Further Adventures of the Political Animal* (Boston: Harcourt Brace Jovanovich, 1987).

22. Raymond Williams, *Keywords: A Vocabulary of Culture and Society* (New York: Oxford University Press, 1976), 184.

23. John Muir, *Nature Writings* (New York: Library of America, 1997), 245.

24. Quoted in Leo Strauss, *Natural Right and History* (Chicago: University of Chicago Press, 1999), 22.

25. Thoreau, *Walking*, 91.

26. Quoted in Charlene Spretnak, *The Resurgence of the Real* (New York: Routledge, 1999), 54.

27. Julian Simon, *The Ultimate Resource* (Princeton, NJ: Princeton University Press, 1981); Julian Simon, *The Ultimate Resource 2* (Princeton, NJ: Princeton University Press, 1996).

28. Scholars analyze environmental politics through various lenses. One of the most helpful is Jennifer Clapp and Peter Dauvergne, *Paths to a Green World: The Political Economy of the Global Environment* (Cambridge, MA: MIT Press, 2005).

29. Quoted in Whiteside, *Divided Natures*, 31.

30. Alan Watts, *The Wisdom of Insecurity* (New York: Vintage Books, 1968).

31. Referenced in Thomas Friedman, *Hot, Flat, and Crowded: Why We Need a Green Revolution—and How It Can Renew America* (New York: Farrar, Straus and Giroux, 2008), 316.

Chapter 2

1. For an insightful rereading of the American environmentalist tradition with a focus on thinkers who did not draw a sharp boundary between humans and nature, see, Ben A. Minteer, *Nature in Common? Environmental Ethics and the Contested Foundations of Environmental Policy* (Philadelphia: Temple University Press, 2009); Ben

A. Minteer, *The Landscape of Reform: Civic Pragmatism and Environmental Thought in America* (Cambridge, MA: MIT Press, 2006); Andrew Light and Eric Katz, *Environmental Pragmatism* (London: Routledge, 1996).

2. Ramachandra Guha, *Environmentalism: A Global History* (New York: Longman, 2000).

3. Henry David Thoreau, *Walking: A Little Book of Wisdom* (New York: HarperCollins, 1994), 34.

4. Thomas Malthus, *An Essay on the Principle of Population* (New York: Oxford University Press, 1999).

5. Rachel Carson, *Silent Spring* (Boston: Houghton Mifflin, 1987), 6.

6. Ibid., 99.

7. Paul Ehrlich, *The Population Bomb* (New York: Balantine Books, 1968), 44.

8. Donella Meadows, Jorgen Randers, Dennis Meadows, and William Behrens III, *The Limits to Growth: A Report for the Club of Rome's Project on the Predicament of Mankind* (New York: Signet, 1972), 134.

9. Donella Meadows, Jorgen Randers, and Dennis Meadows, *Limits to Growth: The 30-Year Update* (White River Junction, VT: Chelsea Green Publishing, 2004).

10. John Maddox, *The Doomsday Syndrome* (New York: McGraw-Hill, 1972), 4.

Chapter 3

1. Charles Reich, *The Greening of America* (New York: Bantam, 1971).

2. John Stuart Mill, *Three Essays on Religion* (Amherst, NY: Prometheus Books, 1998), 11.

3. United Nations Environment Programme, *The Fourth Global Environment Outlook: Environment for Development (GEO-4)* (Valletta, Malta: Progress Press Ltd., 2007), 122.

4. Ibid., 124.

5. James Kunstler, *The Long Emergency: Surviving the Converging Catastrophes of the Twenty-First Century* (New York: Atlantic Monthly Press, 2005), 24–30.

6. Robert Torrance, introduction to *Encompassing Nature: Nature and Culture from Ancient Times to the Modern World*, ed. Robert Torrance (Washington, DC: Counterpoint, 1999), xiv–xv.

7. Quoted in Luc Ferry, *What Is the Good Life?* trans. Lydia G. Cochrane (Chicago: University of Chicago Press, 2005), 156.

8. Barry Commoner, *The Closing Circle: Nature, Man, and Technology*, 1st ed. (New York: Knopf, 1971), 37.

9. Andrew Dobson, *Green Political Thought*, 3rd ed. (New York: Routledge, 2000).

10. See, for example, Murray Bookchin, *The Ecology of Freedom: The Emergence and Dissolution of Hierarchy* (Warner, NH: Writers Publishing Cooperative, 2003).

11. See, for example, Kirkpatrick Sale, *Dwellers in the Land: The Bioregional Vision* (San Francisco: Sierra Club Books, 1985).

12. See, for example, Fritjof Capra, *The Hidden Connections: Integrating the Biological, Cognitive, and Social Dimensions of Life into a Science of Sustainability* (New York: Doubleday, 2002); Paul Hawken, Amory Lovins, and L. Hunter Lovins, *Natural Capitalism: Creating the Next Industrial Revolution* (New York: Back Bay Books, 2000).

13. John Meyer, *Political Nature: Environmentalism and the Interpretation of Western Thought* (Cambridge, MA: MIT Press, 2001).

14. Wendell Berry, *Sex, Economy, Freedom, and Community* (New York: Pantheon 1993), 11.

15. William McDonough and Michael Braungart, *Cradle to Cradle* (New York: North Point Press, 2002).

16. Wes Jackson, *Altars of Unhewn Stone: Science and the Earth* (San Francisco: North Point Press, 1987); Judith Soule and Jon Piper, *Farming in Nature's Image* (Washington, DC: Island Press, 1992).

17. Paul Hawken, *The Ecology of Commerce: A Declaration of Sustainability*, 1st ed. (New York: HarperCollins Publishers, 1993); Ben Cohen and Mal Warwick, *Values-Driven Business: How to Change the World, Make Money, and Have Fun* (San Francisco: Berrett-Koehler Publishers, 2006).

18. Janine Benyus, *Biomimicry: Innovation Inspired by Nature* (New York: Perennial, 2002).

19. See Aristotle *Physics* 2.1–9; Aristotle *Metaphysics* 5.4; Aristotle *On the Heavens* 3.2.

20. Lynn White, "Historical Roots of Our Ecological Crisis," *Science* 155, no. 3767 (March 1967): 1203–1207.

21. Fritjof Capra, *The Web of Life: A New Understanding of Living Systems* (New York: Anchor/Doubleday, 1997); Charlene Spretnak, *The Resurgence of the Real* (New York: Routledge, 1999).

22. Karen Warren, *Ecofeminist Philosophy* (Lanham, MD: Rowman and Littlefield, 2000); Joel Kovel, *The Enemy of Nature: The End of Capitalism or the End of the World* (Boston: Zed Books, 2002); James Gustave Speth, *The Bridge at the Edge of the World: Capitalism, the Environment, and Crossing from Crisis to Sustainability* (New Haven, CT: Yale University Press 2008).

23. Peter Singer, *Animal Liberation: A New Ethics for Our Treatment of Animals* (New York: Random House, 1975).

24. Tom Regan, *The Case for Animal Rights* (Berkeley: University of California Press, 2004).

25. Arne Naess, "The Shallow and the Deep: Long-Range Ecology Movements," *Inquiry* 16 (1973): 95–100.

26. Aldo Leopold, *A Sand County Almanac and Sketches Here and There* (New York: Oxford University Press, 1989), 204.

27. Dave Foreman, *Confessions of an Eco-Warrior* (New York: Three Rivers Press, 1993), 192.

28. See William Cronon, "The Trouble with Wilderness; or Getting Back to the Wrong Nature," in *Uncommon Ground: Rethinking the Human Place in Nature*, ed. William Cronon (New York: W. W. Norton, 1996).

29. John Muir, *Our National Parks* (Boston: Houghton Mifflin, 1901), 1.

30. Richard Louv, *Last Child in the Woods: Saving Our Children from Nature-Deficit Disorder* (Chapel Hill, NC: Algonquin Books, 2005).

31. Henry David Thoreau, *Walden*, ed. Walter Harding (Boston: Houghton Mifflin, 1995), 97.

32. Ralph Waldo Emerson and Brooks Atkinson, *The Essential Writings of Ralph Waldo Emerson* (New York: Modern Library, 2000).

33. John Muir, *The Wilderness World of John Muir*, ed. Edwin Way Teale (New York: Mariner Books, 2001), 70.

34. Philip Shabecoff, *A Fierce Green Fire: The American Environmental Movement* (Washington, DC: Island Press, 2003), 152.

35. EPPN, "The Arctic National Wildlife Refuge and the Episcopal Church, USA: Frequently Asked Questions," in *The Arctic National Wildlife Refuge and the Episcopal Church, USA*, available at <http://www.episcopalchurch.org/3654_36643_ENG_HTM.htm>;   PANGAEA, "Arctic National Wildlife Refuge Poster," in *Arctic National Wildlife Refuge Poster*, available at <http://www.pangaea.org/arctic%20national%20wildlife%20refuge/anwr.htm>.

36. Edward O. Wilson, *Biophilia* (Cambridge, MA: Harvard University Press, 1984).

37. Martin H. Krieger, "What's Wrong with Plastic Trees?" *Science* 179 (February 1973): 446–455.

Chapter 4

1. See Charlene Spretnak, *The Resurgence of the Real* (New York: Routledge, 1999), 45.

2. See, for example, Stephen Moore, "The Coming Age of Abundance," in *The True State of the Planet*, ed. Ronald Bailey (New York: Free Press, 1995).

3. Ronald Bailey, ed., *The True State of the Planet* (New York: Free Press, 1995).

4. Bjørn Lomborg, *The Skeptical Environmentalist: Measuring the Real State of the World* (Cambridge: Cambridge University Press, 2001); Gregg Easterbrook, *A Moment on Earth: The Coming Age of Environmental Optimism* (New York: Penguin, 1996).

5. Harold Barnett and Chandler Morse, *Scarcity and Growth: The Economics of Natural Resource Availability* (Baltimore: Johns Hopkins University Press, 1963).

6. Julian Simon, *The Economics of Population Growth* (Princeton, NJ: Princeton University Press, 1977); Julian Simon, ed., *Research in Population Economics*, 2 vols. (Greenwich, CT: JAI Publishers, 1978).

7. Ester Boserup, *The Conditions of Agricultural Growth: The Economics of Agrarian Change under Population Pressure* (New York: Earthscan Publications, 1992).

8. Julian Simon, *The Ultimate Resource* (Princeton, NJ: Princeton University Press, 1981).

9. See, for example, Patrick Michaels and Robert Balling, Jr. *Climate of Extremes: Global Warming Science They Don't Want You to Know* (Washington, DC: Cato Institute, 2009). For a host of skeptical views on climate change, see Tim Flannery, "Endgame," *New York Review of Books* 52, no. 13 (August 11, 2005): 26–29.

10. Julian Simon, *The Ultimate Resource 2* (Princeton, NJ: Princeton University Press, 1996), 12.

11. James Trefil, *Human Nature: A Blueprint for Managing the Earth—by People, for People* (New York: Henry Holt and Company, 2004); Paul Driessen, *Eco-Imperialism: Green Power, Black Death* (Bellevue, WA: Free Enterprise Press, 2004).

12. Vaclav Klaus, *Blue Planet in Green Shackles; What Is Endangered: Climate or Freedom?* (Washington, DC: Competitive Enterprise Institute, 2007); Wilfred Beckerman, *A Poverty of Reason: Sustainable Development and Economic Growth* (Oakland, CA: Independent Institute, 2003); Indur Goklany, *The Improving State of the World: Why We're Living Longer, Healthier, More Comfortable Lives on a Clean Planet* (Washington, DC: Cato Institute, 2007).

13. Simon Young, *Designer Evolution: A Transhumanist Manifesto* (New York: Prometheus Books, 2006), 40.

14. Ibid., 50.

15. Quoted in Paul Hawken, *Blessed Unrest: How the Largest Social Movement in History Is Restoring Grace, Justice, and Beauty to the World* (New York: Penguin, 2008), 54.

16. John Stuart Mill, *Three Essays on Religion* (Amherst, NY: Prometheus Books, 1998), 29.

17. Quoted in Spretnak, *The Resurgence of the Real*, 54.

18. Sherwin Nuland, *Maimonides* (New York: Schocken, 2005).

19. John Meyer, *Political Nature: Environmentalism and the Interpretation of Western Thought* (Cambridge, MA: MIT Press, 2001).

20. Lee Silver, *Challenging Nature: The Clash of Science and Spirituality at the New Frontiers of Life* (New York: HarperCollins, 2006); Luc Ferry, *The New Ecological Order* (Chicago: University of Chicago Press, 1995).

21. Aristotle, *The Nichomachean Ethics* (New York: Oxford University Press, 1998).

22. Jane Jacobs, *The Death and Life of Great American Cities* (New York: Vintage Books, 1992).

23. See *Phaedo*, in Plato, *The Collected Dialogues of Plato, Including the Letters*, ed. Edith Hamilton and Huntington Cairns (New York: Pantheon Books, 1961).

Chapter 5

1. Harold D. Lasswell, *Politics: Who Gets What, When, How* (New York: McGraw-Hill, 1936).

2. William Cronon, ed., *Uncommon Ground: Rethinking the Human Place in Nature* (New York: W. W. Norton, 1996), 36.

3. John R. McNeill, *Something New under the Sun: An Environmental History of the Twentieth-Century World* (New York: W. W. Norton, 2001).

4. Bill McKibben, *Maybe One: A Personal and Environmental Argument for Single-Child Families* (New York: Simon and Schuster 1998), 85

5. United Nations Environment Programme, *The Fourth Global Environment Outlook: Environment for Development (GEO-4)* (Valletta, Malta: Progress Press Ltd., 2007), 122.

6. Bruce Sterling, quoted in Ross Robertson, "A Brighter Shade of Green: Rebooting Environmentalism for the 21st Century," *What Is Enlightenment?* 38 (2007): 44.

7. Edward O. Wilson, *The Future of Life* (New York: Vintage, 2003), 102.

8. Quoted in Leslie Thiele, *Environmentalism for a New Millennium* (New York: Oxford University Press, 1999), 48–49.

9. Walter Truett Anderson, *To Govern Evolution: Further Adventures of the Political Animal* (Boston: Harcourt Brace Jovanovich, 1987).

10. Tim Flannery, *The Weather Makers* (New York: HarperCollins, 2006).

11. See Bill McKibben, *Wandering Home* (New York: Crown Publishers, 2005), 118.

12. Walter Truett Anderson, *Evolution Isn't What It Used to Be* (San Francisco: W. H. Freeman and Company, 1996).

13. Donna Jeanne Haraway, *Simians, Cyborgs, and Women: The Reinvention of Nature* (New York: Routledge, 1991), 152.

14. Andrew Pollack, "Researchers Announce a Step toward Synthetic Life," *New York Times*, January 25, 2008, A15.

15. James Trefil, *Human Nature: A Blueprint for Managing the Earth—by People, for People* (New York: Henry Holt and Company, 2004), 8.

16. Cited in Bill McKibben, *Enough: Staying Human in an Engineered Age* (New York: Owl Books, 2004), 25.

17. Gregory Stock, *Redesigning Humans: Our Inevitable Genetic Future* (New York: Houghton Mifflin, 2002), 4; see also, Francis Fukuyama, *Our Posthuman Future: Consequences of the Biotechnology Revolution* (New York: Picador, 2003).

18. Sam Harris, *The End of Faith: Religion, Terror, and the Future of Reason* (New York: W. W. Norton, 2005), 220.

19. Andrea Wansbury, *Birds: Divine Messengers: Transform Your Life with Their Guidance and Wisdom* (Findhorn, Scotland: Findhorn Press, 2006).

20. William Cronon, "The Trouble with Wilderness; or Getting Back to the Wrong Nature," in *Uncommon Ground: Rethinking the Human Place in Nature*, ed. William Cronon (New York: W. W. Norton, 1996), 88.

21. Steven Vogel, "Habermas and the Ethics of Nature," in *The Ecological Community*, ed. Roger Gottlieb (New York: Routledge, 1997), 184.

22. Quoted in David Harvey, *Justice, Nature, and the Geography of Difference* (Cambridge, MA: Blackwell, 1996), 26.

Chapter 6

1. J. Baird Callicott, "The Wilderness Idea Revisited: The Sustainable Development Alternative," in *The Great New Wilderness Debate*, ed. J. Baird Callicott and Michael Nelson (Athens: University of Georgia Press, 1998).

2. Stephen J. Pyne, *How the Canyon Became Grand: A Short History* (New York: Penguin Books, 1998).

3. Ibid., 28, 37, 41.

4. Michael David Spence, *Dispossessing the Wilderness: Indian Removal and the Making of the National Parks* (New York: Oxford University Press, 2000).

5. Karl Jacoby, *Crimes against Nature: Squatters, Poachers, Thieves, and the Hidden History of American Conservation* (Berkeley: University of California Press, 2001).

6. Mahesh Rangarajan and Ghazala Shahabuddin, "Displacement and Relocation from Protected Areas: Towards a Biological and Historical Synthesis," *Conservation and Society* 4, no. 3 (2006): 359–378.

7. Robert Hitchcock, "'We Are the First People': Land, Natural Resources, and Identity in the Central Kalahari, Botswana," *Journal of Southern African Studies* 28, no. 4 (2002): 814.

8. Theodore Binnema and Melanie Niemi, "'Let the Line Be Drawn Now': Wilderness, Conservation, and the Exclusion of Aboriginal People from Banff National Park in Canada," *Environmental History* 11, no. 4 (2006): 724–751.

9. *The Wilderness Act of 1964*, Public Law 88-577 (16 U.S. C. 1131-1136), 88th Cong., 2nd sess. (September 3, 1964), sec. 2 (c).

10. Daniel Glick, "Of Lynx and Men," *National Geographic*, January 2006, Vol. 209 (1), 56–67.

11. Tim Rawson, *Changing Tracks: Predators and Politics in Mt. McKinley National Park* (Fairbanks: University of Alaska Press, 2003); Victoria Butler, "Elephants by the Truckload," *International Wildlife* 25, no. 4 (1995): 30–36.

12. Lawrence Downes, "Aloha, Po'ouli: Farewell to a Hawaii Native We Will Never Meet Again," *New York Times*, December 19, 2004, WK10.

13. Jonathan Adams, *The Future of the Wild: Radical Conservation for a Crowded World* (Boston: Beacon, 2006).

14. Arun Agrawal and Kent Redford, "Conservation and Displacement: An Overview," in *Protected Areas and Human Displacement: A Conservation Perspective*, ed. Kent Redford and Eva Fearn (Bronx: Wildlife Conservation Society, 2007), 7.

15. Mark Woods, "Wilderness," in *A Companion to Environmental Philosophy*, ed. Dale Jamieson (Malden, MA: Blackwell Publishing, 2003), 351.

16. Aldo Leopold, *A Sand County Almanac and Sketches Here and There* (New York: Oxford University Press, 1989), 199.

17. Dave Foreman, "Wilderness: From Scenery to Nature," in *Wild Earth: Wild Ideas for a World Out of Balance*, ed. Tom Butler (Minneapolis: Milkweed, 2002), 16–17.

18. John Terborgh, *Requiem for Nature* (Washington, DC: Shearwater Books, 1999).

19. Connie Barlow, "Rewilding for Evolution," in *Wild Earth: Wild Ideas for a World Out of Balance*, ed. Tom Butler (Minneapolis: Milkweed, 2002), 85.

20. Terborgh, *Requiem for Nature*, 63.

21. Michael Rosenzweig, *Win-Win Ecology: How the Earth's Species Can Survive in the Midst of Human Enterprise* (New York: Oxford University Press, 2003); Adams, *The Future of the Wild*.

22. Lee Silver, *Challenging Nature: The Clash of Science and Spirituality at the New Frontiers of Life* (New York: HarperCollins, 2006), 313.

23. Jack Turner, *The Abstract Wild* (Tucson: University of Arizona Press, 1996), 26.

24. Dennis Martinez, "Protected Areas, Indigenous Peoples, and the Western Idea of Nature," *Ecological Restoration* 21, no. 4 (2003): 249.

25. Ghazala Shahabuddin, "Conservation and Ecology at a Crossroads," PowerPoint, Washington, DC, 2008.

26. David Abram, "Why Wild," Alliance for Wild Ethics, available at <http://www.wildethics.org/why_wild.html>.

27. Paul Rezendes, *The Wild Within: Adventures in Nature and Animal Teachings* (New York: Berkley Trade 1999).

28. William Cronon, "The Trouble with Wilderness; or Getting Back to the Wrong Nature," in *Uncommon Ground: Rethinking the Human Place in Nature*, ed. William Cronon (New York: W. W. Norton, 1996), 88.

Chapter 7

1. Bill McKibben, *The End of Nature* (New York: Random House, 1989), 58.

2. Edward O. Wilson, *The Creation: An Appeal to Save Life on Earth* (New York: W. W. Norton, 2006), 15.

3. McKibben, *The End of Nature*, 58.

4. William McDonough and Michael Braungart, *Cradle to Cradle* (New York: North Point Press, 2002), 45.

5. Mark Pagani, Ken Caldeira, David Archer, and James C. Zachos, "An Ancient Carbon Mystery," *Science* 314, no. 5805 (2006): 1556–1557.

6. National Research Council, *Surface Temperature Reconstructions for the Last 2,000 Years* (Washington, DC: National Academy Press, 2006).

7. Intergovernmental Panel on Climate Change, *Climate Change 2007: Synthesis Report* (Valencia, Spain: Intergovernmental Panel on Climate Change, 2007), 37–38.

8. Ted Nordhaus and Michael Shellenberger, *Break Through: From the Death of Environmentalism to the Politics of Possibility* (New York: Houghton Mifflin 2007), 221.

9. Wilfred Beckerman, *A Poverty of Reason: Sustainable Development and Economic Growth* (Oakland, CA: Independent Institute, 2003); Bjørn Lomborg, *The Skeptical Environmentalist: Measuring the Real State of the World* (Cambridge: Cambridge University Press, 2001); Indur Goklany, *The Improving State of the World: Why We're Living Longer, Healthier, More Comfortable Lives on a Cleaner Planet* (Washington, DC: Cato Institute, 2007); Bjørn Lomborg, *Cool It: The Skeptical Environmentalist's Guide to Global Warming*, 1st ed. (New York: Alfred A. Knopf, 2007).

10. Lomborg, *Cool It*, 39.

11. McDonough and Braungart, *Cradle to Cradle*, 54.

12. Bill McKibben, *Deep Economy: The Wealth of Communities and the Durable Future* (New York: Times Books, 2007), 183–184. See also Thomas Friedman, *Hot, Flat, and Crowded: Why We Need a Green Revolution—and How It Can Renew America* (New York: Farrar, Straus and Giroux, 2008), 73.

13. Quoted in Patrick C. Kangas, *Ecological Engineering: Principles and Practice* (Boca Raton, FL: Lewis Publishers, 2004), 314.

14. Ken Buessler, Scott Doney, David Karl, Philip Boyd, and Ken Caldeira, "Ocean Iron Fertilization: Moving Forward in a Sea of Uncertainty," *Science* 319, no. 5860 (2008), 162; Robert Kunzig, "Pick up a Mop," *Time*, July 14, 40-43.

15. Paul Crutzen, "Albedo Enhancement by Stratospheric Sulfur Injections: A Contribution to Resolve a Policy Dilemma?" *Climatic Change* 77, no. 3–4 (2006): 211–220.

16. Molly Bentley, "Guns and Sunshades to Rescue Climate," *BBC News*, 2007, available at <http://news.bbc.co.uk/2/hi/science/nature/4762720.stm>.

17. Robert Angel, "Feasibility of Cooling the Earth with a Cloud of Small Spacecraft near the Inner Lagrange Point (L1)," *Proceedings*

of the National Academy of Sciences of the United States of America 103, no. 46 (2006): 17184–17189.

18. Robin McKie and Juliette Jowit, "Can Science Really Save the World?" *Observer* (England) (2007): 22.

19. Jonathan Leak, "Sounds Crazy But It Might Save the Planet," *Sunday Times* (March 18, 2007), 6.

20. McDonough and Braungart, *Cradle to Cradle*, 30.

21. Worldwatch Institute and Center for American Progress, ed., *American Energy: The Renewable Path to Energy Security* (Washington, DC: Worldwatch Institute, 2006), 10.

22. Jay Inslee and Bracken Hendricks, *Apollo's Fire: Igniting American's Clean Energy Economy* (Washington, DC: Island Press, 2007), 18.

23. Stephen Pacala and Robert Socolow, "Stabilization Wedges: Solving the Climate Problem for the Next 50 Years with Current Technologies," *Science* 305, no. 5686 (2004): 968–972; Inslee and Hendricks, *Apollo's Fire.*

Chapter 8

1. Kerry Whiteside, *Divided Natures: French Contributions to Political Ecology* (Cambridge, MA: MIT Press, 2002), 261.

2. Quoted in Bill Plotkin, *Nature and the Human Soul: Cultivating Wholeness and Community in a Fragmented World* (Novato, CA: New World Library, 2008), 149.

3. Francis Fukuyama, *Our Posthuman Future: Consequences of the Biotechnology Revolution* (New York: Picador, 2003).

4. Michael Shellenberger and Ted Nordhaus, *Break Through: From the Death of Environmentalism to the Politics of Possibility* (New York: Houghton Mifflin, 2007); William McDonough and Michael Braungart, *Cradle to Cradle* (New York: North Point Press, 2002); Thomas L. Friedman, *Hot, Flat, and Crowded: Why We Need a Green Revolution—and How It Can Renew America* (New York: Farrar, Straus and Giroux, 2008); Jay Inslee and Bracken Hendricks, *Apollo's Fire: Igniting American's Clean Energy Economy* (Washington, DC: Island Press, 2007); Alex Steffen, *Worldchanging: A User's Guide for the 21st Century* (New York: Abrams, 2006).

5. See, for example, Bill McKibben, "Green Fantasia," *New York Review of Books 55*, no. 17 (2008); Joseph Romm, "Debunking Shellenberger and Nordhaus: Parts I–IV," *Grist Environmental News and Commentary*, available at <http://www.grist.org/article/debunking-shellenberger-nordhaus-part-i>.

6. See, for example, Bryan G. Norton, *Sustainability: A Philosophy of Adaptive Ecosystem Management* (Chicago: University of Chicago Press, 2005); Kerry Whiteside, *Precautionary Politics: Principle and Practice in Confronting Environmental Risk* (Cambridge, MA: MIT Press, 2006); David Orr, *The Nature of Design: Ecology, Culture, and Human Intention* (New York: Oxford University Press, 2002).

7. Michael J. Sandel, *The Case against Perfection: Ethics in the Age of Genetic Engineering* (Cambridge, MA: Belknap Press, 2007).

8. Aldo Leopold, *A Sand County Almanac and Sketches Here and There* (New York: Oxford University Press, 1989), 197.

References

Abram, David. 2009. Alliance for Wild Ethics. Available at: http://www.wildethics.org/

Adams, Jonathan. 2006. *The Future of the Wild: Radical Conservation for a Crowded World*. Boston: Beacon.

Agrawal, Arun, and Kent Redford. 2007. Conservation and Displacement: An Overview. In *Protected Areas and Human Displacement: A Conservation Perspective*, ed. K. Redford and E. Fearn. Bronx: Wildlife Conservation Society.

Anderson, Walter Truett. 1987. *To Govern Evolution: Further Adventures of the Political Animal*. Boston: Harcourt Brace Jovanovich.

Anderson, Walter Truett. 1996. *Evolution Isn't What It Used To Be*. San Francisco: W. H. Freeman and Company.

Angel, Robert. 2006. Feasibility of Cooling the Earth with a Cloud of Small Spacecraft near the Inner Lagrange Point (L1). Proceedings of the National Academy of Sciences of the United States of America 103 (46): 17184–17189.

Aristotle. 1998. *The Nichomachean Ethics*. New York: Oxford University Press.

Bailey, Ronald, ed. 1995. *The True State of the Planet*. New York: Free Press.

Barlow, Connie. 2002. Rewilding for Evolution. In *Wild Earth: Wild Ideas for a World Out of Balance*, ed. T. Butler. Minneapolis: Milkweed.

Barnett, Harold, and Chandler Morse. 1963. *Scarcity and Growth: The Economics of Natural Resource Availability*. Baltimore: Johns Hopkins University Press.

Beckerman, Wilfred. 2003. *A Poverty of Reason: Sustainable Development and Economic Growth*. Oakland, CA: Independent Institute.

Bentley, Molly. 2007. Guns and Sunshades to Rescue Climate. *BBC News*. Available from <http://news.bbc.co.uk/2/hi/science/nature/4762720.stm>.

Benyus, Janine. 2002. *Biomimicry: Innovation Inspired by Nature*. New York: Perennial.

Berry, Wendell. 1994. *Sex, Economy, Freedom and Community: Eight Essays*. New York: Pantheon.

Binnema, Theodore, and Melanie Niemi. 2006. "Let the Line Be Drawn Now": Wilderness, Conservation, and the Exclusion of Aboriginal People from Banff National Park in Canada. Environmental History 11 (4): 724–751.

Bookchin, Murray. 2003. *The Ecology of Freedom: The Emergence and Dissolution of Hierarchy*. Warner, NH: Writers Publishing Cooperative.

Borgmann, Albert. 1995. The Nature of Reality and the Reality of Nature. In *Reinventing Nature? Responses to Postmodern Deconstruction*, ed. M. E. Soule and G. Lease. Washington, DC: Island Press.

Boserup, Ester. 1992. *The Conditions of Agricultural Growth: The Economics of Agrarian Change under Population Pressure*. New York: Earthscan Publications.

Botkin, Daniel. 1992. *Discordant Harmonies: A New Ecology for the Twenty-first Century*. New York: Oxford University Press.

Buesseler, Ken, Scott Doney, David Karl, Philip Boyd, and Ken Caldeira. 2008. Ocean Iron Fertilization: Moving Forward in a Sea of Uncertainty. Science 319 (5860): 162.

Butler, Victoria. 1995. Elephants by the Truckload. International Wildlife 25 (4): 30–36.

Callicott, J. Baird. 1998. The Wilderness Idea Revisited: The Sustainable Development Alternative. In *The Great New Wilderness Debate*, ed. J. B. Callicott and M. Nelson. Athens: University of Georgia Press.

Capra, Fritjof. 1997. *The Web of Life: A New Understanding of Living Systems*. New York: Anchor/Doubleday.

Capra, Fritjof. 2002. *The Hidden Connections: Integrating the Biological, Cognitive, and Social Dimensions of Life into a Science of Sustainability*. New York: Doubleday.

Carson, Rachel. 1987. *Silent Spring.* Boston: Houghton Mifflin.

Clapp, Jennifer and Peter Dauvergne. 2005. *Paths to a Green World: The Political Economy of the Global Environment.* Cambridge, MA: MIT Press.

Cohen, Ben, and Mal Warwick. 2006. *Values-Driven Business: How to Change the World, Make Money, and Have Fun.* San Francisco: Berrett-Koehler Publishers.

Commoner, Barry. 1971. *The Closing Circle: Nature, Man, and Technology.* 1st ed. New York: Knopf.

Cronon, William. 1996a. The Trouble with Wilderness; or Getting Back to the Wrong Nature. In *Uncommon Ground: Rethinking the Human Place in Nature,* ed. W. Cronon. New York: W. W. Norton.

Cronon, William, ed. 1996b. *Uncommon Ground: Rethinking the Human Place in Nature.* New York: W. W. Norton.

Crutzen, Paul. 2006. Albedo Enhancement by Stratospheric Sulfur Injections: A Contribution to Resolve a Policy Dilemma? Climatic Change 77 (3–4): 211–220.

Descartes, René. 1999. *Discourse on Method and Meditations on First Philosophy.* Trans. D. A. Cress. 4 ed. Indianapolis: Hackett.

Dobson, Andrew. 2000. *Green Political Thought.* 3rd ed. New York: Routledge.

Downes, Lawrence. 2004. Aloha, Po'ouli: Farewell to a Hawaii Native We Will Never Meet Again. *New York Times,* December 19, WK10.

Driessen, Paul. 2004. *Eco-Imperialism: Green Power, Black Death.* Bellevue, WA: Free Enterprise Press.

Easterbrook, Gregg. 1996. *A Moment on Earth: The Coming Age of Environmental Optimism.* New York: Penguin.

Ehrenfeld, David W. 1981. *The Arrogance of Humanism.* New York: Oxford University Press.

Ehrlich, Paul. 1968. *The Population Bomb.* New York: Balantine Books.

Emerson, Ralph Waldo, and Brooks Atkinson. 2000. *The Essential Writings of Ralph Waldo Emerson.* New York: Modern Library.

Ferry, Luc. 1995. *The New Ecological Order.* Chicago: University of Chicago Press.

Ferry, Luc. 2005. *What Is the Good Life?* Translated by L. G. Cochrane. Chicago: University of Chicago Press.

Fischer, Frank, and Maarten Hajer, eds. 1999. *Living with Nature: Environmental Politics as Cultural Discourse.* New York: Oxford University Press.

Flannery, Tim. 2005. Endgame. *New York Review of Books* 52, no. 13 (August 11): 26–29.

Flannery, Tim. 2006. *The Weather Makers.* New York: HarperCollins.

Foreman, Dave. 1993. *Confessions of an Eco-Warrior.* New York: Three Rivers Press.

Foreman, Dave. 2002. Wilderness: From Scenery to Nature. In *Wild Earth: Wild Ideas for a World Out of Balance,* ed. T. Butler. Minneapolis: Milkweed.

Friedman, Thomas. 2008. *Hot, Flat, and Crowded: Why We Need a Green Revolution—and How It Can Renew America.* New York: Farrar, Straus and Giroux.

Fukuyama, Francis. 2003. *Our Posthuman Future: Consequences of the Biotechnology Revolution.* New York: Picador.

Glick, Daniel. 2006. Of Lynx and Men. National Geographic 209 (1), (January): 56–67.

Goklany, Indur. 2007. *The Improving State of the World: Why We're Living Longer, Healthier, More Comfortable Lives on a Clean Planet.* Washington, DC: Cato Institute.

Gore, Albert. 2007. *The Assault on Reason.* New York: Penguin Press.

Guha, Ramachandra. 2000. *Environmentalism: A Global History.* New York: Longman.

Haraway, Donna Jeanne. 1991. *Simians, Cyborgs, and Women: The Reinvention of Nature.* New York: Routledge.

Harris, Sam. 2005. *The End of Faith: Religion, Terror, and the Future of Reason.* New York: W. W. Norton.

Harvey, David. 1996. *Justice, Nature, and the Geography of Difference.* Cambridge, MA: Blackwell.

Hawken, Paul. 1993. *The Ecology of Commerce: A Declaration of Sustainability.* 1st ed. New York: HarperCollins Publishers.

Hawken, Paul. 2008. *Blessed Unrest: How the Largest Social Movement in History Is Restoring Grace, Justice, and Beauty to the World.* New York: Penguin.

Hawken, Paul, Amory Lovins, and L. Hunter Lovins. 2000. *Natural Capitalism: Creating the Next Industrial Revolution*. New York: Back Bay Books.

Hitchcock, Robert. 2002. "We Are the First People": Land, Natural Resources, and Identity in the Central Kalahari, Botswana. Journal of Southern African Studies 28 (4): 797–824.

Hull, R. Bruce. 2006. *Infinite Nature*. Chicago: University of Chicago Press.

Inslee, Jay, and Bracken Hendricks. 2007. *Apollo's Fire: Igniting American's Clean Energy Economy*. Washington, DC: Island Press.

Intergovernmental Panel on Climate Change. 2007. *Climate Change 2007: Synthesis Report*. Valencia, Spain: Intergovernmental Panel on Climate Change.

Jackson, Wes. 1987. *Altars of Unhewn Stone: Science and the Earth*. San Francisco: North Point Press.

Jacobs, Jane. 1992. *The Death and Life of Great American Cities*. New York: Vintage Books.

Jacoby, Karl. 2001. *Crimes against Nature: Squatters, Poachers, Thieves, and the Hidden History of American Conservation*. Berkeley: University of California Press.

Kangas, Patrick C. 2004. *Ecological Engineering: Principles and Practice*. Boca Raton, FL: Lewis Publishers.

Klaus, Vaclav. 2007. *Blue Planet in Green Shackles; What Is Endangered: Climate or Freedom?* Washington, DC: Competitive Enterprise Institute.

Kovel, Joel. 2002. *The Enemy of Nature: The End of Capitalism or the End of the World*. Boston: Zed Books.

Krieger, Martin H. 1973. What's Wrong with Plastic Trees? *Science* 179: 446–455.

Kunstler, James. 2005. *The Long Emergency: Surviving the Converging Catastrophes of the Twenty-First Century*. New York: Atlantic Monthly Press.

Kunzig, Robert. 2008. Pick Up a Mop. *Time Magazine* (July 14), 40–43.

Lasswell, Harold D. 1936. *Politics: Who Gets What, When, How*. New York: McGraw-Hill.

Leak, Jonathan. 2007. Sounds Crazy But It Might Save the Planet. Sunday Times *(London)* (March 18): 6.

Leopold, Aldo. 1989. *A Sand County Almanac and Sketches Here and There.* New York: Oxford University Press.

Light, Andrew, and Eric Katz. 1996. *Environmental Pragmatism.* London: Routledge.

Lomborg, Bjørn. 2001. *The Skeptical Environmentalist: Measuring the Real State of the World.* Cambridge: Cambridge University Press.

Lomborg, Bjørn. 2007. *Cool It: The Skeptical Environmentalist's Guide to Global Warming.* 1st ed. New York: Alfred A. Knopf.

Louv, Richard. 2005. *Last Child in the Woods: Saving Our Children from Nature-Deficit Disorder.* Chapel Hill, NC: Algonquin Books.

Luke, Timothy. 1997. *Ecocritique: Contesting the Politics of Nature, Economy, and Culture.* Minneapolis: University of Minnesota Press.

Maddox, John. 1972. *The Doomsday Syndrome.* New York: McGraw-Hill.

Malthus, Thomas. 1999. *An Essay on the Principle of Population.* New York: Oxford University Press.

Marsh, George Perkins. 2006. *Man and Nature; or, Physical Geography as Modified by Human Action.* Ann Arbor: Scholarly Publishing Office, University of Michigan.

Martinez, Dennis. 2003. Protected Areas, Indigenous Peoples, and the Western Idea of Nature. Ecological Research 21 (4): 247–250.

McDonough, William, and Michael Braungart. 2002. *Cradle to Cradle.* New York: North Point Press.

McKibben, Bill. 1989. *The End of Nature.* New York: Random House.

McKibben, Bill. 1991. *Maybe One: A Case for Smaller Families.* New York: Plume.

McKibben, Bill. 2004. *Enough: Staying Human in an Engineered Age.* New York: Owl Books.

McKibben, Bill. 2005. *Wandering Home.* New York: Crown Publishers.

McKibben, Bill. 2007. *Deep Economy: The Wealth of Communities and the Durable Future.* New York: Times Books.

McKibben, Bill. 2008. Green Fantasia. *New York Review of Books* 55 (17).

McKie, Robin, and Juliette Jowit. 2007. Can Science Really Save the World? *Observer* (London), 22.

McNeill, John R. 2001. *Something New under the Sun: An Environmental History of the Twentieth-Century World*. New York: W. W. Norton.

Meadows, Donella, Jorgen Randers, and Dennis Meadows. 2004. *Limits to Growth: The 30-Year Update*. White River Junction, VT: Chelsea Green Publishing.

Meadows, Donella, Jorgen Randers, Dennis Meadows, and William Behrens III. 1972. *The Limits to Growth: A Report for the Club of Rome's Project on the Predicament of Mankind*. New York: Signet.

Meyer, John. 2001. *Political Nature: Environmentalism and the Interpretation of Western Thought*. Cambridge, MA: MIT Press.

Michaels, Patrick, and Robert Balling, Jr. 2009. *Climate of Extremes: Global Warming Science They Don't Want You to Know*. Washington, DC: Cato Institute.

Mill, John Stuart. 1998. *Three Essays on Religion*. Amherst, NY: Prometheus Books.

Minteer, Ben A. 2006. *The Landscape of Reform: Civic Pragmatism and Environmental Thought in America*. Cambridge, MA: MIT Press.

Minteer, Ben A. 2009. *Nature in Common? Environmental Ethics and the Contested Foundations of Environmental Policy*. Philadelphia: Temple University Press.

Moore, Stephen. 1995. *The Coming Age of Abundance*. Ed. Ronald Bailey. New York: Free Press.

Muir, John. 1901. *Our National Parks*. Boston: Houghton Mifflin.

Muir, John. 1997. *Nature Writings*. New York: Library of America.

Muir, John. 2001. *The Wilderness World of John Muir*. Ed. E. W. Teale. New York: Mariner Books.

Naess, Arne. 1973. The Shallow and the Deep: Long-Range Ecology Movements. Inquiry 16:95–100.

National Research Council. 2006. *Surface Temperature Reconstructions for the Last 2,000 Years*. Washington, DC: National Academy Press.

Nordhaus, Ted, and Michael Shellenberger. 2007. *Break Through: From the Death of Environmentalism to the Politics of Possibility.* New York: Houghton Mifflin.

Norton, Bryan G. 2005. *Sustainability: A Philosophy of Adaptive Ecosystem Management.* Chicago: University of Chicago Press.

Noss, Reed F. 2002. Wilderness—Now More Than Ever: A Response to Callicott. In *Wild Earth: Wild Ideas for a World Out of Balance,* ed. T. Butler. Minneapolis: Milkweed Editions.

Nuland, Sherwin. 2005. *Maimonides.* New York: Schocken.

Orr, David. 2002. *The Nature of Design: Ecology, Culture, and Human Intention.* New York: Oxford University Press.

Pacala, Stephen, and Socolow, Robert 2004. Stabilization Wedges: Solving the Climate Problem for the Next 50 Years with Current Technologies. *Science* 305 (5686): 968–972.

Pagani, Mark, Ken Caldeira, David Archer, and James C. Zachos. 2006. An Ancient Carbon Mystery. *Science* 314 (5805): 1556–1557.

Parris, Matthew. 2008. Nature Superior to Man? What Green Twaddle. *TimesOnline.* August 30. Available at <http://www.timesonline.co.uk/tol/comment/columnists/matthew_parris/article4636286.ece>.

Plato. 1961. *The Collected Dialogues of Plato, Including the Letters.* Ed. Edith Hamilton and Huntington Cairns. New York: Pantheon Books.

Plotkin, Bill. 2008. *Nature and the Human Soul: Cultivating Wholeness and Community in a Fragmented World.* Novato, CA: New World Library.

Pollack, Andrew. 2008. Researchers Announce a Step toward Synthetic Life, *New York Times,* January 25, A15.

Pollan, Michael. 1991. *Second Nature: A Gardener's Education.* New York: Bantam.

Pope, Carl. 2004. *Strategic Ignorance: Why the Bush Administration Is Recklessly Destroying a Century of Environmental Progress.* San Francisco: Sierra Club Books.

Pyne, Stephen J. 1998. *How the Canyon Became Grand: A Short History.* New York: Penguin Books.

Rangarajan, Mahesh, and Ghazala Shahabuddin. 2006. Displacement and Relocation from Protected Areas: Towards a Biological and Historical Synthesis. *Conservation and Society* 4 (3): 359–378.

Rawson, Tim. 2003. *Changing Tracks: Predators and Politics in Mt. McKinley National Park*. Fairbanks: University of Alaska Press.

Regan, Tom. 2004. *The Case for Animal Rights*. Berkeley: University of California Press.

Reich, Charles. 1971. *The Greening of America*. New York: Bantam.

Rezendes, Paul. 1999. *The Wild Within: Adventures in Nature and Animal Teachings*. New York: Berkley Trade.

Rilke, Rainer Maria. 1981. *Selected Poems of Rainer Maria Rilke*. Translated by R. Bly. New York: Harper and Row.

Robertson, Ross. 2007. A Brighter Shade of Green: Rebooting Environmentalism for the 21st Century. *What Is Enlightenment?* 38: 44.

Rolston, Holmes, III. 1998. The Wilderness Idea Reaffirmed. In *The Great New Wilderness Debate*, ed. J. B. Callicott and M. Nelson. Athens: University of Georgia Press.

Romm, Joseph. 2008. Debunking Shellenberger and Nordhaus: Parts I-IV. *Grist Environmental News and Commentary*. Available at http://www.grist.org/article/debunking-shellenberger-nordhaus-part-i.

Rosenzweig, Michael. 2003. *Win-Win Ecology: How the Earth's Species Can Survive in the Midst of Human Enterprise*. New York: Oxford University Press.

Sale, Kirkpatrick. 1985. *Dwellers in the Land: The Bioregional Vision*. San Francisco: Sierra Club Books.

Sandel, Michael J. 2007. *The Case against Perfection: Ethics in the Age of Genetic Engineering*. Cambridge, MA: Belknap Press.

Shabecoff, Philip. 2003. *A Fierce Green Fire: The American Environmental Movement*. Washington, DC: Island Press.

Shahabuddin, Ghazala. 2008. *Conservation and Ecology at a Crossroads*. Washington, DC: PowerPoint.

Shellenberger, Michael, and Ted Nordhaus. 2007. *Break Through: From the Death of Environmentalism to the Politics of Possibility*. New York: Houghton Mifflin.

Silver, Lee. 2006. *Challenging Nature: The Clash of Science and Spirituality at the New Frontiers of Life*. New York: HarperCollins.

Simon, Julian. 1977. *The Economics of Population Growth*. Princeton, NJ: Princeton University Press.

Simon, Julian, ed. 1978. *Research in Population Economics. 2 vols.* Greenwich, CT: JAI Publishers.

Simon, Julian. 1981. *The Ultimate Resource*. Princeton, NJ: Princeton University Press.

Simon, Julian. 1996. *The Ultimate Resource 2*. Princeton, NJ: Princeton University Press.

Singer, Peter. 1975. *Animal Liberation: A New Ethics for Our Treatment of Animals*. New York: Random House.

Snyder, Gary. 2002. Is Nature Real? In *Wild Earth: Wild Ideas for a World Out of Balance*, ed. T. Butler. Minneapolis: Milkweed Editions.

Sokal, Alan, and Jean Bricmont. 1999. *Fashionable Nonsense: Postmodern Intellectuals' Abuse of Science*. 1st ed. New York: Picador.

Soper, Kate. 1995. *What Is Nature? Culture, Politics, and the Non-Human*. London: Blackwell.

Soule, Judith, and Jon Piper. 1992. *Farming in Nature's Image*. Washington, DC: Island Press.

Spence, Michael David. 2000. *Dispossessing the Wilderness: Indian Removal and the Making of the National Parks*. New York: Oxford University Press.

Speth, James Gustave. 2008. *The Bridge at the Edge of the World: Capitalism, the Environment, and Crossing from Crisis to Sustainability*. New Haven, CT: Yale University Press.

Spretnak, Charlene. 1999. *The Resurgence of the Real*. New York: Routledge.

Steffen, Alex. 2006. *Worldchanging: A User's Guide for the 21st Century*. New York: Abrams.

Stock, Gregory. 2002. *Redesigning Humans: Our Inevitable Genetic Future*. New York: Houghton Mifflin.

Strauss, Leo. 1999. *Natural Right and History*. Chicago: University of Chicago Press.

Terborgh, John. 1999. *Requiem for Nature*. Washington, DC: Shearwater Books.

Thiele, Leslie. 1999. *Environmentalism for a New Millennium*. New York: Oxford University Press.

Thoreau, Henry David. 1994. *Walking: A Little Book of Wisdom*. New York: HarperCollins.

Thoreau, Henry David. 1995. *Walden*. Ed. W. Harding. Boston: Houghton Mifflin.

Torrance, Robert. 1999. Introduction to *Encompassing Nature: Nature and Culture from Ancient Times to the Modern World*, ed. R. Torrance. Washington, DC: Counterpoint.

Trefil, James. 2004. *Human Nature: A Blueprint for Managing the Earth—by People, for People*. New York: Henry Holt and Company.

Turner, Jack. 1996. *The Abstract Wild*. Tucson: University of Arizona Press.

United Nations Environment Programme. 2007. *The Fourth Global Environment Outlook: Environment for Development (GEO-4)*. Valletta, Malta: Progress Press Ltd.

Vogel, Steven. 1997. Habermas and the Ethics of Nature. In *The Ecological Community*, ed. R. Gottlieb. New York: Routledge.

Waller, Donald. 1996–1997. Wilderness Redux. *Wild Earth* 6 (4): 36–45.

Wansbury, Andrea. 2006. *Birds: Divine Messengers: Transform Your Life with Their Guidance and Wisdom*. Findhorn, Scotland: Findhorn Press.

Wapner, Paul. 1996. *Environmental Activism and World Civic Politics*. Albany: State University of New York.

Warren, Karen. 2000. *Ecofeminist Philosophy*. Lanham, MD: Rowman and Littlefield.

Watts, Alan. 1968. *The Wisdom of Insecurity*. New York: Vintage Books.

White, Lynn. 1967. Historical Roots of Our Ecological Crisis. *Science* 155 (3767): 1203–1207.

Whiteside, Kerry. 2002. *Divided Natures: French Contributions to Political Ecology*. Cambridge, MA: MIT Press.

Whiteside, Kerry. 2006. *Precautionary Politics: Principle and Practice in Confronting Environmental Risk*. Cambridge, MA: MIT Press.

Williams, Raymond. 1976. *Keywords: A Vocabulary of Culture and Society*. New York: Oxford University Press.

Wilson, Edward O. 1984. *Biophilia*. Cambridge, MA: Harvard University Press.

Wilson, Edward O. 2003. *The Future of Life*. New York: Vintage.

Wilson, Edward O. 2006. *The Creation: An Appeal to Save Life on Earth*. New York: W. W. Norton.

Woods, Mark. 2003. Wilderness. In *A Companion to Environmental Philosophy*, ed. D. Jamieson. Malden, MA: Blackwell Publishing..

Worldwatch Institute and Center for American Progress, ed. 2006. *American Energy: The Renewable Path to Energy Security*. Washington, DC: Worldwatch Institute.

Young, Oran. 2002. *The Institutional Dimensions of Environmental Change: Fit, Interplay, and Scale*. Cambridge, MA: MIT Press.

Young, Simon. 2006. *Designer Evolution: A Transhumanist Manifesto*. New York: Prometheus Books.

Index